Signals and Communication Technology

This series is devoted to fundamentals and applications of modern methods of signal processing and cutting-edge communication technologies. The main topics are information and signal theory, acoustical signal processing, image processing and multimedia systems, mobile and wireless communications, and computer and communication networks. Volumes in the series address researchers in academia and industrial R&D departments. The series is application-oriented. The level of presentation of each individual volume, however, depends on the subject and can range from practical to scientific.

Indexing: All books in "Signals and Communication Technology" are indexed by Scopus and zbMATH

For general information about this book series, comments or suggestions, please contact Mary James at mary.james@springer.com or Ramesh Nath Premnath at ramesh.premnath@springer.com.

Mourad Abbas
Editor

Practical Solutions for Diverse Real-World NLP Applications

Springer

Editor
Mourad Abbas
High Council of Arabic
Algiers, Algeria

ISSN 1860-4862 ISSN 1860-4870 (electronic)
Signals and Communication Technology
ISBN 978-3-031-44259-9 ISBN 978-3-031-44260-5 (eBook)
https://doi.org/10.1007/978-3-031-44260-5

This Springer imprint is published by the registered company Springer Nature Switzerland AG
The registered company address is: Gewerbestrasse 11, 6330 Cham, Switzerland

Preface

This book provides a range of research that presents new ideas to resolve issues regarding miscellaneous applications related to NLP. It covers topics such as data augmentation, phoneme recognition in low-resource languages, error correction detection, identification and aggregation of duplicate security findings, and exploiting social media discourse to estimate typical affective responses to music.

As data is a critical component in training high-performing deep learning models, this book begins with presenting techniques of text augmentation and paraphrase generation. It then discusses, for the sake of a better readability of texts, the use of generative byte-level transformer models for restoring spaces, punctuation, and capitalization in different languages. This is an important step useful for many applications as post-processing of noisy data and ASR outputs. The book also proposes a multi-task learning approach for multilingual phoneme recognition in low-resource languages.

Dialog systems are one of the major widespread applications, but they can sometimes be prone to errors and ambiguities; that is why a chapter in this book is devoted to error correction detection and error correction, leading to disambiguation.

The book also shows how NLP presents solutions to predict music emotion values from social media comments without analysing audio signal or processing lyrics information. Finally, it discusses how to identify and aggregate the duplicate findings in the industry.

We hope this book will help the readers to gain a deeper understanding of the covered topics and apply the information to their own research.

Algiers, Algeria | Mourad Abbas

Contents

Probabilistic Linguistic Knowledge and Token-Level Text Augmentation

Zhengxiang Wang

1 Introduction

Data serves as a crucial component in training high-performing and robust machine learning models that can effectively tackle real-world learning tasks. However, data availability is often unpredictable and not guaranteed. In the realm of supervised learning, the development of reliably deployable models typically requires the collection of vast amounts of annotated data, which is affordable only for a select few. In low-resource settings, in particular, the available data may be limited or entirely nonexistent. There are also situations where existing data are imbalanced for specific classes, causing models trained on such data to be easily biased toward classes with abundant training examples. This can potentially be harmful when the models are deployed. Practical considerations like these have given rise to data augmentation, a widely adopted strategy to mitigate the problems of scarce or imbalanced data. Data augmentation involves applying label-preserving transformations to existing data to generate novel labeled data. This approach has seen considerable success in various fields, such as image and speech recognition [1–7].

Text augmentation, a subcategory of data augmentation that focuses on augmenting text data, is a promising yet challenging domain within NLP [8–13]. The challenge arises due to the lack of well-established methods that can consistently generate diverse and accurate augmented texts simultaneously. In contrast to images or speech, where physical features can be relatively easily manipulated without altering the label, text is highly sensitive to the arrangement and combination of words. For instance, to augment an image, one can rotate, crop, flip, or change

Z. Wang (✉)
Department of Linguistics, Stony Brook University, Stony Brook, NY, USA
e-mail: zhengxiang.wang@stonybrook.edu

M. Abbas (ed.), *Practical Solutions for Diverse Real-World NLP Applications*,
Signals and Communication Technology,
https://doi.org/10.1007/978-3-031-44260-5_1

its color specifications in a predetermined manner, while still assuming that the augmented images represent the same object as the original [1]. However, when augmenting text, one cannot merely replace and shuffle words in an automated fashion to generate paraphrases. It becomes evident that there is a need for foundational research exploring the factors that influence the effectiveness of text augmentation [9].

The primary objective of this study is to gain a deeper understanding of the effectiveness of token-level text augmentation within a linguistically motivated evaluation context. Token-level text augmentation is assessed, as opposed to other more complex methods (see Sect. 2), due to its applicability across various tasks and languages. The insights regarding its effectiveness may prove valuable in low-resource domains, where text augmentation is primarily employed in real-world scenarios. More specifically, this study aims to address the following two research questions: (1) How effective is token-level text augmentation? (2) Can the incorporation of probabilistic linguistic knowledge enhance the effectiveness of text augmentation?

To address these two research questions, comprehensive experiments were conducted on a binary question matching classification task, involving both Chinese and English languages, at a reasonably large scale. The objective of this task is to predict whether two given questions share the same expressed intent. This type of task, which entails the classification of text pairs, is well suited for evaluating text augmentation as it demands high fidelity of the augmented texts to the original texts to maintain label preservation. Conversely, since token-level text augmentation is not strictly paraphrastic, its success in such tasks serves as strong evidence of its overall effectiveness. To explore the impact of probabilistic linguistic knowledge, pretrained n-gram language models are also utilized to select augmented texts, which, in theory, should be statistically more likely and expected to be closer to natural texts or of higher quality. Consequently, it is anticipated that probabilistic linguistic knowledge will enhance the effectiveness of text augmentation.

The chapter proceeds as follows: Sect. 2 reviews related works, while the augmentation methods and experimental settings of the study are detailed in Sects. 3 and 4, respectively. Section 5 presents the results of the main experiments, and the findings of three supplementary experiments are reported in Sect. 6. Section 7 offers further discussion on the discoveries and provides a conclusion.

2 Related Works

Over the years, three major types of text augmentation have been employed in NLP to generate label-preserving data [13]: token-level augmentation, sentence-level augmentation, and hidden-level augmentation. Token-level augmentation focuses on individual tokens and involves word replacements, which can be either dictionary-based [14] or embedding-based [15], as well as deletion, insertion, or shuffling of tokens in a random [16] or predefined [17, 18] manner. Sentence-level augmentation, on the other hand, typically entails paraphrasing at the sentence level. Back translation [19–21], a widely popular technique that involves translating a text into another language and then retranslating it back, exemplifies this approach. Additionally, the researchers have utilized language models to generate novel data conditioned on given text or labels [22–24]. Lastly, hidden-level augmentation pertains to the manipulation of hidden representations, such as linear transformations, in order to create new perturbed representations [25–27].

Many of the aforementioned studies have reported slight yet inconsistent performance gains when training models with augmented data for their respective NLP tasks, mostly text classification tasks. A common explanation for any observed performance improvement is that the augmented data introduces noise to the original training data, thus preventing the trained models from overfitting [28]. This, in turn, improves their performance on test sets.

A notable and widely cited example is provided by [16]. The paper employs four simple token-level text editing operations to augment train sets of varying sizes and demonstrates their general effectiveness in boosting model performance across five sentiment-related and text type classification tasks. Although claimed to be universal, these four text editing operations, which are also examined in this study (see Sect. 3), have been found not to be consistently beneficial. More specifically, they have been shown to negatively impact model performance in more complex tasks, such as natural language inference and paraphrasing [13], and fail to consistently improve performance for transformers [29].

This study aims to enrich the existing literature on the effectiveness of token-level text augmentation by conducting comprehensive and fine-grained cross-linguistic experiments for an under-explored task. It additionally examines the role of probabilistic linguistic knowledge, also an under-explored yet fundamental question.

3 Augmentation Methods

3.1 Text Augmentation Techniques

In this chapter, five token-level text augmentation techniques, or text editing operations, are employed: Synonym Replacement (SR), Random Swap (RS), Random Insertion (RI), Random Deletion (RD), and Random Mix (RM).

The first four techniques were initially proposed by [16] as a simple but universal set of text augmentation techniques named as EDA (Easy Data Augmentation). For a single text edit, they work as follows. SR randomly replaces a word, where possible, with one of its randomly sampled synonyms (if more than one) based on a predefined dictionary. RS, on the other hand, swaps the positions of a random word pair within a text. RI inserts a random synonym immediately after an eligible target word, while RD deletes a word at random. For n text edits, these techniques are simply applied n times. Additionally, RM, introduced in [30], is a random combination of 2–4 of the other four techniques, resulting in a more diversified text. Given their randomness, these techniques are also referred to as random text perturbations [31].

The five text augmentation techniques are implemented in Python using a program called REDA (Revised EDA). In addition to incorporating an extra text editing operation (RM), REDA differs from EDA in three key aspects. Firstly, REDA prevents duplicates in the output text(s), which can occur when there are no synonyms available for replacement (SR) or insertion (RS) for words in the input text or when the same words are replaced or swapped back during the SR and RS operations. Secondly, REDA does not preprocess the input text (e.g., removing stop words), as this is believed to have minimal impact and better aligns with the fundamental concept of random text perturbations that underlie these augmentation techniques. Lastly, REDA replaces only one word with its synonym at a given position per text edit, rather than replacing all occurrences, which are regarded as additional edits.

In this chapter, the synonym dictionary for English is derived from WordNet [32], while for Chinese, it is obtained from multiple reputable sources through web scraping.[1] Furthermore, rather than applying a fixed number of text edits to texts of varying lengths, this study employs an *editing rate*, which adjusts the number of text edits proportionally to the text length.

3.2 N-Gram Language Model

A n-gram is a string of n words $W_1^n = (w_1, \dots w_n)$. A n-gram language model is a Markov model that estimates the probability (typically expressed in logarithmic terms [33]) of a given string W_1^L of length L (where $L \geq n$) by the product of the probabilities of all its n-long substrings.

$$\log P(W_i^L) \approx \log \prod_{i=1}^{L-n+1} P(W_i^{i+n-1}) = \sum_{i=1}^{L-n+1} \log P(W_i^{i+n-1}) \tag{1}$$

where $P(W_i^{i+n-1})$ represents $P(w_{i+n-1}|W_i^{i+n-2})$. Let $C(W_i^{i+n-1})$ denote the frequency of occurrence of a string W_i^{i+n-1} in a much larger corpus used as the training data. The maximum-likelihood probability estimate for $P(W_i^{i+n-1})$ is the relative frequency of W_i^{i+n-1} against its previous $(n-1)$ words W_i^{i+n-2} in counts:

$$P(W_i^{i+n-1}) \approx \frac{C(W_i^{i+n-1})}{C(W_i^{i+n-2})} \tag{2}$$

[1] https://github.com/jaaack-wang/Chinese-Synonyms.

Since both $C(W_i^{i+n-1})$ and $C(W_i^{i+n-2})$ can be 0 during the deployment of a pretrained n-gram language model, this leads to inaccurate or undefined probability estimates. Inspired by both Eq. (1) and Stupid Backoff [34], this study uses a non-discounting method that estimates the probability of an unseen n-gram by multiplying the probabilities of its two $(n-1)$-grams together, as shown in Eq. (3). The method will continue to back off into unigrams if all other higher order n-grams do not occur in the training data, where unseen unigrams are simply assigned the same probability as those one-off unigrams.

$$P(W_i^{i+n-1}) \approx \begin{cases} \frac{C(W_i^{i+n-1})}{C(W_i^{i+n-2})}, & \text{if } C(W_i^{i+n-1}) > 0 \\ \prod_{i=1}^{2} P(W_i^{i+n-2}), & \text{otherwise} \end{cases} \tag{3}$$

In this chapter, I trained both the Chinese and English n-gram language models using n-grams up to 4-grams. The pretrained n-gram language models are utilized as a filter to select the k most likely outputs from m possible outputs generated by REDA, where m is at least 20 times greater than k in this study. The program that combines REDA with an n-gram language model is denoted as REDA_{NG}.

4 Experimental Settings

4.1 Task and Data

The task under analysis is a binary classification task on the basis of text pairs, commonly known as question matching. The aim is to predict whether a given question pair (Q, Q') expresses similar intents, which is a fundamental sub-task for question answering or, more broadly, a downstream task for semantic matching [35]. Two labels are used, with 0 denoting negative match of (Q, Q') and 1 positive match.

This chapter considers two large-scale benchmark datasets for question matching. One is the Large-scale Chinese Question Matching Corpus (LCQMC, [35]) for Chinese. The other is the Quora Question Pairs Dataset (QQQD)[2] for English.

For LCQMC, I reused the original train, development, and test sets as provided by [35]. For QQQD, three label-balanced datasets were created from its train set since the test set is made unlabeled for online competition. The basic statistics about these two datasets are given in Table 1.

[2] https://quoradata.quora.com/First-Quora-Dataset-Release-Question-Pairs.

Table 1 Statistics of the data splits for LCQMC and QQQD

Split	LCQMC (matched and mismatched)	QQQD (matched and mismatched)
Train	238,766 (138,574 & 100,192)	260,000 (130,000 & 130,000)
Dev	8802 (4402 & 4400)	20,000 (10,000 & 10,000)
Test	12,500 (6250 & 6250)	18,526 (9263 & 9263)

4.2 Classification Models

In the main experiments, four classic neural network models were chosen: Continuous Bag of Words (CBOW, [36]), Convolutional Neural Network (CNN, [37]), Gated Recurrent Units (GRU, [38]), and Long Short-Term Memory (LSTM, [39]). Since the focus here is to evaluate the effectiveness of token-level text augmentation and the role of probabilistic linguistic knowledge, the use of various classification models is not meant to contrast the learning difference among them, but rather to make the examination more comprehensive. Pretrained word embeddings were not utilized to simulate low-resource settings. For the same reason, transformers [40] were only used in the supplementary experiments, instead of the main ones.

The models can also be divided into three groups, depending on the type of train sets they train on. *Baseline models* refer to models that train on train sets without augmentation, i.e., train sets that only contain original training examples. Models that train on train sets augmented by REDA are called as *REDA models*, and similarly *$REDA_{NG}$ models* are the ones that train on train sets augmented by $REDA_{NG}$. REDA models and $REDA_{NG}$ models are also called *augmented models*, since both are trained on augmented train sets. Augmented train sets contain augmented examples on the top of the original examples, based on which the augmented examples are produced. For convenience, the three respective types of train sets are simply called *baseline train sets*, *REDA train sets*, and *$REDA_{NG}$ train sets*.

4.3 Training Details

The models were constructed in PaddlePaddle,[3] a deep learning framework developed by Baidu and trained on Baidu Machine Learning CodeLab's AI Studio with Tesla V100 GPU and 32G RAM. The models were trained using mini batches of size

[3] https://www.paddlepaddle.org.cn/en.

64 with the objective of reducing cross-entropy loss. The Adam optimizer [41] with 5e-4 learning rate was applied to assist training. The training time was consistently 3 epochs long, since most of the models overfitted the train sets within 3 epochs. Development sets were used for validation purposes.

The basic structure for the classification models is simple and unified as follows. Each model begins with two separate embedding layers with the same embedding size to convert the input text pair (Q, Q') into two respective embedded sequences, $\mathbf{Embd}_Q$ and $\mathbf{Embd}_{Q'}$. Then, $\mathbf{Embd}_Q$ and $\mathbf{Embd}_{Q'}$ each pass through an encoder layer, whose structure is determined by the classification model in use, to obtain two encoded sequences, $\mathbf{Enc}_Q$ and $\mathbf{Enc}_{Q'}$. The encoded sequences, $\mathbf{Enc}_Q$ and $\mathbf{Enc}_{Q'}$, are concatenated along the last axis and then passed to a fully connected feed-forward network (FFN) that consists of two linear transformations with a tanh activation function in between.

$$FFN(\mathbf{x}) = tanh(\mathbf{x}\mathbf{W}_1 + b_1)\mathbf{W}_2 + b_2 \tag{4}$$

For CBOW, the encoder is the point-wise summation of the embeddings of tokens in the input text pair followed a tanh function. For the rest, the encoder layers are simply CNN layer, GRU layer, and LSTM layer, corresponding to the model names.

4.4 Augmentation Details

Due to experimental costs, it was not possible for this study to evaluate the effects of different initializations of REDA/REDA$_{NG}$ (i.e., editing rate, number of augmentation per example) on the trained models' performance. Therefore, I initialized REDA/REDA$_{NG}$ with small editing rates, informed by the authors in [16], who recommend small editing rates over large ones and demonstrate that large editing rates lead to performance decline in their ablation experiments. This makes sense since large editing rates are more likely to cause label changes. Intuitively, if small editing rates do not work well, larger ones will not either. The number of augmentation per example, to be mentioned below, was kept small for the same consideration.

More concretely, REDA and REDA$_{NG}$ were initialized with the following editing rates for SR, RS, RI, and RD, respectively, 0.2, 0.2, 0.1, and 0.1. I applied Python rounding rule to calculate and perform the number of edits needed for each operation. That means, if the number of edits is less than or equal to 0.5, it will be rounded down to 0, and thus no editing operation will apply. To make the experiments more controlled and doable, (1) I made RM only randomly perform two of the other four editing operations with one edit each, and (2) every editing operation produced up to 2 non-duplicated augmented texts per text (or 4 per text pair), if the train set size was less than 50k; otherwise, there would only be one augmented text per text instead. Every augmented text was crossed-paired with the other text that was the pair to the text being augmented, with the original

Table 2 Size of augmented train sets for the main experiments on LCQMC and QQQD. For convenience, 240k is hereafter used to refer to the full size (i.e., 238,766 to be exact) of LCQMC. Note that all the subsets of the full train sets were randomly and independently sampled

LCQMC	Augmented		QQQD	Augmented
5k	66,267		10k	148,341
10k	132,513		50k	543,066
50k	563,228		100k	1,086,063
100k	929,176		150k	1,629,178
240k	2,218,512		260k	2,823,733

label retained for the augmented text pair. These settings were also applied for the supplementary experiments.

Table 2 shows the size of the augmented train sets for the main experiments.

5 Main Experiments

This section reports the test set performance of the four classification models trained on train sets of varying size with and without augmentation for the binary question matching task in Chinese and English. As the test sets for LCQMC and QQQD are equally balanced across labels (see Sect. 4.1), accuracy is considered as the primary evaluation metric. The average precision and recall are taken as secondary metrics for more nuanced analyses.

5.1 *Chinese: LCQMC*

Table 3 shows the test set accuracy of the four classification models trained on the three types of train sets (baseline, REDA, and REDA_{NG}) of different sizes. Contrary to the expectation, incorporating probabilistic linguistic knowledge into the five token-level text augmentation techniques does not lead to superior model performance, as the REDA_{NG} models never outperform the REDA models in terms of average performance, when given the same amounts of training data. Instead, the REDA models almost always achieve slightly better performance than the REDA_{NG} counterparts. Moreover, it appears that the augmented models (REDA and REDA_{NG}) do not necessarily have better test set accuracy than that of the baseline models, unless augmentation is applied to sufficient original training examples (i.e., at least 50k).

The average test set precision and recall, as shown in Table 4, may elucidate the performance gains of over the baseline models. There are two factors at play.

Table 3 Test set accuracy (%) of the four classification models trained on LCQMC's train sets of varying size with and without augmentation. The header denotes train set sizes in terms of original training examples. The best performance given a train set size and a model type is highlighted in **bold**

Models	5k	10k	50k	100k	240k
CBOW	**59.4**	60.4	65.4	67.8	73.8
+REDA	58.1	**60.9**	**68.2**	**72.2**	**76.4**
+REDA$_{NG}$	58.8	59.6	68.1	71.2	76.0
CNN	59.3	**63.4**	67.2	69.0	72.9
+REDA	59.8	62.6	66.8	**69.8**	**74.9**
+REDA$_{NG}$	**60.3**	62.0	**67.9**	69.1	74.0
LSTM	**60.0**	**62.1**	66.2	69.6	74.8
+REDA	58.9	61.5	**67.7**	**71.8**	**76.4**
+REDA$_{NG}$	57.7	60.9	**67.7**	71.7	75.9
GRU	**59.8**	**61.9**	68.1	70.3	**76.8**
+REDA	58.7	61.3	**68.7**	**72.7**	**76.8**
+REDA$_{NG}$	58.8	60.0	67.8	72.5	76.6
Average	**59.6**	**62.0**	66.7	69.2	74.6
+REDA	58.9	61.6	**67.9**	**71.6**	**76.1**
+REDA$_{NG}$	58.9	60.6	**67.9**	71.1	75.6

Table 4 Average test set precision and recall (%) of the four classification models trained on LCQMC's train sets of varying size with and without augmentation. The best performance given a train set size and a metric is highlighted in **bold**

Models	5k	10k	50k	100k	240k
Precision	**57.4**	59.2	62.8	64.3	69.0
+REDA	56.8	**59.5**	64.1	**66.8**	**70.3**
+REDA$_{NG}$	**57.4**	58.2	**64.4**	66.4	69.9
Recall	**75.2**	**77.5**	**82.0**	**86.2**	89.2
+REDA	73.8	72.7	81.2	85.8	**90.4**
+REDA$_{NG}$	71.1	76.1	79.9	85.5	89.8

First, the augmented models consistently exhibit higher precision than the baseline models starting from the training size 50k. Second, the gap in recall between the baseline models and the augmented models becomes much narrower in favor of the augmented models at the same time. This suggests that the augmented models seem to learn to make significantly fewer false negatives with sufficient original training examples augmented, resulting in a sudden improvement in recall compared to the baseline models. It appears that 50k is a threshold, prior to which augmentation seems detrimental to model performance, despite the substantial increase in training examples.

5.2 English: QQQD

The test set accuracy on QQQD shown in Table 5 exhibits a similar pattern to that on LCQMC for two reasons. First, the difference between REDA and REDA$_{NG}$ models remains negligible, reaffirming the trivial role probabilistic linguistic knowledge plays in the five token-level text augmentation techniques. Second, the augmented models do not outperform the baseline models until a sufficient number of original training examples are seen. However, unlike the experiment on LCQMC, this time the REDA$_{NG}$ models consistently perform better than the REDA models. Moreover, it appears that the threshold for the REDA and REDA$_{NG}$ models to outperform the baseline models is much larger, or 100k and 150k, respectively. These two differences are likely attributable to some training artifacts related to the datasets, the pretrained n-gram language models, and the likes.

Also differing from the LCQMC experiment is how the baseline models compare to the augmented models in terms of average test set precision and recall (see Table 6). Instead of displaying a consistent advantage over the augmented models in one of these two metrics, the baseline models show better precision in a way highly correlated with accuracy. In other words, the augmented models do not outperform the baseline models until their respective thresholds. This indicates that the baseline models tend to make fewer false positives, compared to the augmented models when the original training data is insufficient for the text augmentation to become effective.

Table 5 Test set accuracy (%) of the four classification models trained on QQQD's train sets of varying size with and without augmentation

Models	10k	50k	100k	150k	260k
CBOW	**64.4**	**69.9**	72.1	74.2	77.7
+REDA	62.5	68.5	71.6	74.8	78.0
+REDA$_{NG}$	62.9	69.4	**74.0**	**75.5**	**78.2**
CNN	**66.1**	**71.1**	72.6	73.4	75.9
+REDA	63.7	69.9	**72.7**	**75.3**	77.6
+REDA$_{NG}$	63.5	69.3	**72.7**	74.7	**77.7**
LSTM	**65.7**	**71.6**	**72.9**	75.0	77.9
+REDA	64.0	69.8	72.5	**75.1**	**78.1**
+REDA$_{NG}$	64.9	70.3	72.7	75.0	**78.1**
GRU	**67.2**	**71.0**	**74.3**	74.7	77.4
+REDA	63.3	70.0	72.8	74.8	78.1
+REDA$_{NG}$	64.0	70.2	73.8	**75.7**	**78.9**
Average	**65.9**	**70.9**	73.0	74.3	77.2
+REDA	63.4	69.6	72.4	75.0	78.0
+REDA$_{NG}$	63.8	69.8	**73.3**	**75.2**	**78.2**

Table 6 Average test set precision and recall (%) of the four classification models trained on QQQD's train sets of varying size with and without augmentation

Models	10k	50k	100k	150k	260k
Precision	**65.0**	**69.8**	71.6	72.1	76.0
+REDA	61.8	68.0	70.6	73.6	76.2
+REDA$_{NG}$	62.6	69.0	**72.3**	**74.2**	**77.3**
Recall	69.5	73.6	76.2	**79.4**	79.6
+REDA	**70.4**	**73.7**	**77.0**	78.0	**81.4**
+REDA$_{NG}$	69.2	72.0	75.7	77.4	80.0

5.3 *Interim Summary*

Overall, the results presented above demonstrate that incorporating probabilistic linguistic knowledge into REDA does not make a significant difference. Pairwise Mann–Whitney U tests confirm that there is no statistically significant difference in the test set performance between the REDA and REDA$_{NG}$ models, with the obtained p-values close to 1.0, regardless of the specific metric in use.

Additionally, it is revealed that the five token-level text augmentation techniques are not always effective, irrespective of whether an n-gram language model is employed to optimize the augmented outputs or not. The results indicate that for both Chinese and English binary question matching tasks, the augmented models only outperform the baseline models when a sufficient number of original training examples are augmented.

There are two differences observed between the experiments in Chinese and English. The differences concern the relative performance of the REDA$_{NG}$ models against the REDA models and the relative performance of the augmented models against the baseline models. Nevertheless, these differences do not impact the two general observations made above for the purpose of this study.

6 Supplementary Experiments

Following the main results obtained in Sect. 5, three important follow-up questions arise: (1) Does REDA$_{NG}$ truly produce higher quality augmented texts than REDA under the same conditions? (2) Would the results remain valid if state-of-the-art transformer models were employed instead? (3) What if the five token-level text augmentation techniques were applied separately, rather than together? Question (1) is crucial because it determines whether the insignificant difference between the REDA and REDA$_{NG}$ models found in Sect. 5 is due to the marginal role of probabilistic linguistic knowledge or simply because the texts augmented by REDA and REDA$_{NG}$ are indistinguishable in terms of quality. Questions (2) and (3) assess the generality of the observations made so far.

Due to resource constraints and for simplicity, the supplementary experiments in this section are based on LCQMC.

6.1 Comparison of Texts Augmented by REDA and REDA$_{NG}$

Directly comparing the texts augmented by REDA and REDA$_{NG}$ is not feasible, and three text restoration experiments were therefore designed to approximate the comparison. These experiments assess the ability of both programs to restore natural texts when given distorted texts or a pseudo-synonym dictionary for the following text editing operations: Synonym Replacement (SR), Random Swap (RS), and Random Deletion (RD). Random Insertion (RI) and Random Mix (RM) are omitted since inserting random synonyms is generally not representative of natural language use, and the text quality resulting from RM can be inferred from the other basic operations.

Table 7 presents the average accuracy, with the experiment details provided in Appendix A. As shown, while the performance for both approaches declines as the number of edits increases, REDA$_{NG}$ consistently outperforms REDA. For REDA, restoring the distorted texts to their original form is merely a matter of chance, equal to the reciprocal of the number of possible augmented outputs. However, REDA$_{NG}$ augments texts based on the maximum likelihood principle, which tends to be closer to natural texts. This also holds true when natural texts are used as inputs. For instance, through manual inspection, I found that REDA$_{NG}$ performed much better in selecting appropriate synonyms, a problem for REDA due to its randomness and the ubiquitous existence of polysemy. By measuring the bigram overlap rate and Levenshtein edit distances of output texts randomly swapped twice from the natural texts, I further found that the average overlap rate for REDA was much lower (i.e., 0.29 versus 0.77) and that the average edit distances were much larger (i.e., 3.0 versus 1.4) than REDA$_{NG}$. This suggests that REDA$_{NG}$ preserves more collocational features of natural texts than REDA and thus augments higher quality texts.

Table 7 Average accuracy (%) in three text restoration tasks based on a different number of edits (header). *SR*: Synonym Replacement; *RS*: Random Swap; *RD*: Random Deletion. The best performance given an edit number and an augmentation method is highlighted in **bold**

		One	Two	Three
SR	REDA	22	6	2
	REDA$_{NG}$	**88**	**79**	**64**
RS	REDA	9	4	4
	REDA$_{NG}$	**69**	**41**	**34**
RD	REDA	16	5	2
	REDA$_{NG}$	**39**	**22**	**15**

Table 8 Test set, accuracy, precision, and recall (%) for the Ernie-Gram models fine-tuned on LCQMC's train sets of varying size with and without augmentation

Models	5k	10k	50k	100k	240k
Accuracy	**78.7**	**81.7**	**85.9**	**87.1**	**87.4**
+REDA	77.5	80.3	84.1	85.0	85.7
+REDA$_{NG}$	78.6	80.1	83.8	84.6	85.8
Precision	**71.2**	**74.7**	**80.3**	**81.7**	**82.0**
+REDA	70.0	72.9	77.7	79.1	79.5
+REDA$_{NG}$	71.0	73.2	77.4	78.6	80.0
Recall	**96.6**	95.9	95.2	**95.6**	95.9
+REDA	96.4	**96.4**	**95.6**	95.1	**96.1**
+REDA$_{NG}$	95.7	95.1	95.3	95.1	95.5

6.2 *Effect of Transformer*

ERNIE-Gram [42] is a transformer-based pretrained large language model and was chosen for its state-of-the-art performance on LCQMC. The fine-tuning of the ERNIE-Gram models shared identical training details with the main experiments, except that a smaller learning rate (i.e., 5e-5) was used. Table 8 shows the test set performance across the three metrics for the fine-tuned ERNIE-Gram models.

Not surprisingly, the fine-tuned ERNIE-Gram models achieve significantly better results than the four classification models trained in the main experiments on LCQMC. Notably, using only 5k original examples, the fine-tuned ERNIE-Gram models outperform any model trained in the main experiments, regardless of augmentation. This highlights the impressive effectiveness of transfer learning resulting from fine-tuning large language model on downstream tasks. The implication may be that transfer learning is a more robust and effective way of boosting model performance than the text augmentation approaches considered in this study. However, it remains unknown whether this is also the case in low-resource settings.

Despite the noticeable performance gain, the ERNIE-Gram models fine-tuned on augmented train sets are consistently outperformed by the baseline models without augmentation in terms of both accuracy and precision in the test set. Thus, both text augmentation approaches appear to be overall detrimental to model performance. Furthermore, no evidence indicates a significant difference between REDA and REDA$_{NG}$, even when a transformer, such as ERNIE-Gram, is used.

6.3 *Effect of Single Augmentation Technique*

To understand the role of each text augmentation technique, models were trained on train sets augmented using only one augmentation technique. The original train set was partitioned into 11 different sizes, rather than 5, to validate the observation in

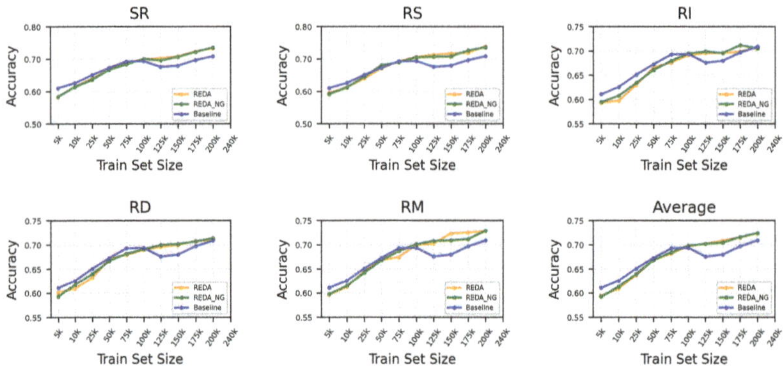

Fig. 1 Average test set accuracy of the four classification models trained on LCQMC's train sets of varying size with and without augmentation under different conditions (i.e., augmentation technique, train size). The sixth plot averages the statistics of the previous five plots

Sect. 5 that the effectiveness of the augmentation is restricted to a sufficient number of original training examples. The experimental details can be found in Appendix B.

Figure 1 displays the average test set accuracy of the four classification models trained on the three types of train sets under different text augmentation techniques and across various training sizes. In line with the previous findings, the effect of probabilistic linguistic knowledge on each of the five techniques is minimal and shows no statistically significant difference, both individually and on average. Also consistent with the previous findings is the existence of a threshold where the augmented models outperform the baseline models in test set accuracy, which appears to be around the training size 100k, rather than 50k as in the related main experiments. The discrepancy may be explained by the different epoch numbers (see Appendix B) and, more importantly, the separation of the augmentation techniques, which, however, are beyond the scope of this chapter.

The average test set precision and recall resemble the data patterns observed in the main experiments, with an updated threshold mentioned above. Please refer to Appendix B for details.

7 Discussion and Conclusion

In this chapter, I evaluate the effectiveness of five token-level text augmentation techniques and the role of probabilistic linguistic knowledge thereof. To this end, two related programs were created: REDA and REDA_{NG}, the latter of which utilizes pretrained n-gram language models to select most likely augmented texts from REDA's output. Experiments on binary question matching classification task in Chinese and English strongly indicate that the role of probabilistic linguistic knowledge for token-level text augmentation is minimal and that the related augmentation techniques are not generally effective. These two findings are further discussed as follows.

First, the difference between the REDA models and the REDA$_{NG}$ models is trivial. However, the supplementary experiment on three pseudo-text restoration tasks in Sect. 6.1 shows that REDA$_{NG}$ arguably generates higher quality augmented texts compared to REDA, as it preserves more collocational features of natural texts. An intuitively plausible explanation for the insignificant role of probabilistic linguistic knowledge may be due to the inherent inability of the five augmentation techniques to produce strictly paraphrastic augmented texts. In other words, the texts augmented by REDA and REDA$_{NG}$ are to a considerable extent comparable in the sense that they are mostly not the paraphrases of the original texts being augmented. Although the REDA$_{NG}$ models appear to be slightly better than the REDA models in the English experiments, the opposite is true for the Chinese experiments. The observed differences are highly likely to result from training artifacts. Nevertheless, none of them are statistically significant.

Second, the five token-level augmentation techniques, whether applied together or separately, are only effective when a sufficiently large number of original training examples are fed into the classification models, including a transformer model. This finding shows that the effectiveness is task-specific and not always positive, contrasting with [16] and aligning with [13, 29]. Unlike the one-text-one-label classification tasks experimented in [16], question matching involves classifying a given question pair into a label to indicate the intentional similarity of the pair. As such, the task is inherently more sensitive to the semantic changes caused by text augmentation and thus arguably represents a more reliable evaluative task. The performance decline in the augmented models in cases with insufficient original training examples may be due to the negative effects of the false matching augmented text pairs generated by REDA and REDA$_{NG}$. However, with enough original training examples seen, the augmented models learn to mediate these negative effects and turn them somewhat into regularization, which helps the models generalize better. Nevertheless, the requirement of a sufficiently large number of training examples makes token-level text augmentation investigated here a less practical and preferable approach for tasks of similar nature to question matching.

One might argue that the differences between REDA/REDA$_{NG}$ and EDA [16], as described in Sect. 3, could be a possible cause for the failure of text augmentation on small train sets in this study. Specifically, by disallowing deduplicates, REDA and REDA$_{NG}$ are more likely to produce more diverse yet non-paraphrastic augmented texts than EDA, given comparably small editing rates. This might exacerbate the negative effects of random text perturbations, thereby requiring more original training examples to mitigate such effects. However, I argue that the differences between REDA/REDA$_{NG}$ and EDA are not crucial. Since the augmented models in this study do not necessarily outperform the baseline models even with a nontrivial number of original training examples (i.e., at least 50k) being augmented, there is no reason to believe that the augmented examples would perform better with fewer original training examples while having the same proportion of augmented examples. Furthermore, it is not surprising that EDA works for simple one-text-one-label classification tasks, despite producing imperfect augmented texts. The reason is exactly task-specific. For example, in a sentence-level sentiment analysis, the sentiment of a sentence is often captured by only few keywords [43]. It follows, as long as an augmented text retains these few keywords or similar replacements, it still reasonably preserves the sentiment label of the original text even if it is grammatically problematic. The key lesson here is that token-level text augmentation may easily introduce noise to the training examples for those simple classification tasks while not causing label changes. As a result, the trained models generalize better.

Systematically and fairly evaluating a text augmentation is uneasy or even unknown. The limitations of this study are obvious, since it fails to experiment with different initializations of REDA/REDA$_{NG}$ or different configurations of the classification models, confined by available computing resources. Nevertheless, this study showcases a linguistically motivated way of evaluating text augmentation and highlights the benefits and insights it provides. The main takeaway is that although token-level text augmentation is simple and potentially useful, it should be used with caution, particularly for complex tasks.

Acknowledgments The chapter is based on two of my previous publications [30, 31]. I thank anonymous reviewers from ICNLSP 2022, ACL 2022 Workshop on Insights from Negative Results in NLP, and AACL-IJCNLP 2022 Workshop on Evaluation & Comparison of NLP Systems, for their feedback. Any remaining errors are solely my responsibility.

Appendix

A. Text Restoration Experiments

For the Synonym Replacement (SR) experiment, I created a pseudo synonym dictionary consisting of 3855 one-word-four-synonym pairs. Each word was mapped to four pseudo synonyms, including the word itself and three non-synonym random words. All the words in the dictionary were those with frequencies ranking between the 1000th and the 10,000th positions in the unigram dictionary compiled for the Chinese n-gram language model. For the Random Swap (RS) and Random Deletion (RD) experiments, I randomly reordered the natural texts and added random words from the texts before performing RS and RD, respectively. For each comparison made, I randomly sampled 10,000 texts from LCQMC's train set for five runs.

B. Ablation Experiments on LCQMC

The training conditions were the same as the main experiments, except for training time. Specifically, to save resources, the training time was reduced to 2 epochs when the train size was 50k or 100k and to 1 epoch when the size was over 100k. Since the aim is to compare the test set performance among the baseline, REDA, and $REDA_{NG}$ models, and because larger train sizes require fewer epochs to fit the train sets, the reduction of the training time is considered reasonable.

For the ablation experiments, Table 9 displays the size of the augmented train sets, and Figs. 2 and 3 show the average test set precision and recall, respectively.

Table 9 Size of augmented train sets per text augmentation technique for LCQMC

Size	SR	RS	RI	RD	RM
5k	24,402	24,758	16,733	16,780	24,859
10k	48,807	49,575	33,090	33,208	49,652
25k	122,358	124,040	83,329	83,592	124,237
50k	244,577	248,074	166,839	167,296	248,539
75k	220,843	223,497	162,563	162,972	224,026
100k	294,516	297,987	216,540	217,012	298,620
125k	368,078	372,536	270,957	271,552	373,266
150k	441,643	446,941	325,027	325,738	447,838
175k	515,229	521,484	379,352	380,214	522,535
200k	588,901	595,977	433,521	434,469	597,084
240k	703,077	711,631	517,492	518,664	712,852

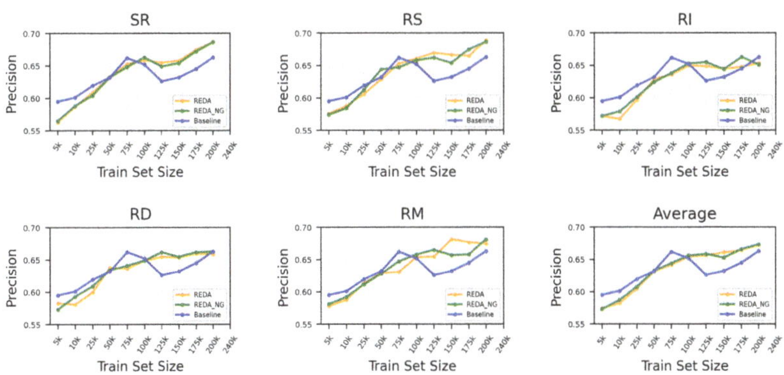

Fig. 2 Average test set precision of the four classification models trained on LCQMC's train sets of varying size with and without augmentation under different conditions

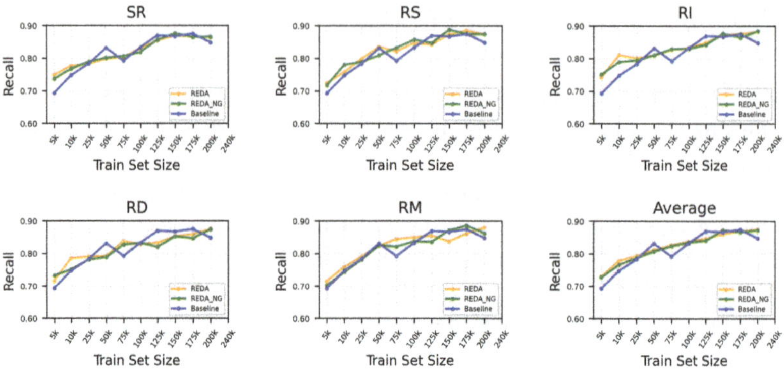

Fig. 3 Average test set recall of the four classification models trained on LCQMC's train sets of varying size with and without augmentation under different conditions

References

1. Simard, P., Steinkraus, D., Platt, J.: Best practices for convolutional neural networks applied to visual document analysis. In: Seventh International Conference On Document Analysis And Recognition, 2003. Proceedings, pp. 958–963 (2003)
2. Krizhevsky, A., Sutskever, I., Hinton, G.: ImageNet classification with deep convolutional neural networks. Adv. Neural Inf. Process. Syst. **25**, (2012)
3. Ko, T., Peddinti, V., Povey, D., Khudanpur, S.: Audio augmentation for speech recognition. In: Proceedings of Interspeech 2015, pp. 3586–3589 (2015)
4. Cui, X., Goel, V., Kingsbury, B.: Data augmentation for deep neural network acoustic modeling. IEEE/ACM Trans. Audio Speech Lang. Process. **23**, 1469–1477 (2015)
5. Park, D., Chan, W., Zhang, Y., Chiu, C., Zoph, B., Cubuk, E., Le, Q.: SpecAugment: A simple data augmentation method for automatic speech recognition. In: Proceedings of Interspeech 2019, pp. 2613–2617 (2019)
6. Shorten, C., Khoshgoftaar, T.: A survey on image data augmentation for deep learning. J. Big Data **6**, 1–48 (2019)
7. Iwana, S.: An empirical survey of data augmentation for time series classification with neural networks. PLOS ONE **16**, 1–32 (2021). https://doi.org/10.1371/journal.pone.0254841
8. Shorten, C., Khoshgoftaar, T., Furht, B.: Text data augmentation for deep learning. J. Big Data **8**, 1–34 (2021)
9. Feng, S., Gangal, V., Wei, J., Chandar, S., Vosoughi, S., Mitamura, T., Hovy, E.: A survey of data augmentation approaches for NLP. In: Findings of the Association for Computational Linguistics: ACL-IJCNLP 2021, pp. 968–988 (2021). https://aclanthology.org/2021.findings-acl.84
10. Liu, P., Wang, X., Xiang, C., Meng, W.: A survey of text data augmentation. In: 2020 International Conference on Computer Communication and Network Security (CCNS), pp. 191–195 (2020)
11. Yang, D., Parikh, A., Raffel, C.: Learning with limited text data. In: Proceedings of the 60th Annual Meeting of the Association for Computational Linguistics: Tutorial Abstracts, pp. 28–31 (2022). https://aclanthology.org/2022.acl-tutorials.5
12. Şahin, G.: To augment or not to augment? A comparative study on text augmentation techniques for low-resource NLP. Comput. Linguist. **48**, 5–42 (2022). https://aclanthology.org/2022.cl-1.2

13. Chen, J., Tam, D., Raffel, C., Bansal, M., Yang, D.: An empirical survey of data augmentation for limited data learning in NLP. Trans. Assoc. Comput. Linguist. **11**, 191–211 (2023). https://doi.org/10.1162/tacl%5C_a%5C_00542
14. Zhang, X., Zhao, J., LeCun, Y.: Character-level convolutional networks for text classification. Adv. Neural Inf. Process. Syst. **28**, (2015). https://proceedings.neurips.cc/paper/2015/file/250cf8b51c773f3f8dc8b4be867a9a02-Paper.pdf
15. Wang, W., Yang, D.: That's so annoying!!!: A lexical and frame-semantic embedding based data augmentation approach to automatic categorization of annoying behaviors using #petpeeve tweets. In: Proceedings of the 2015 Conference on Empirical Methods in Natural Language Processing, pp. 2557–2563 (2015). https://aclanthology.org/D15-1306
16. Wei, J., Zou, K.: EDA: easy data augmentation techniques for boosting performance on text classification tasks. In: Proceedings of the 2019 Conference on Empirical Methods in Natural Language Processing and the 9th International Joint Conference on Natural Language Processing (EMNLP-IJCNLP), pp. 6382–6388 (2019). https://aclanthology.org/D19-1670
17. Kang, D., Khot, T., Sabharwal, A., Hovy, E.: AdvEntuRe: adversarial training for textual entailment with knowledge-guided examples. In: Proceedings of the 56th Annual Meeting of the Association for Computational Linguistics (Volume 1: Long Papers), pp. 2418–2428 (2018). https://aclanthology.org/P18-1225
18. Asai, A., Hajishirzi, H.: Logic-guided data augmentation and regularization for consistent question answering. In: Proceedings of the 58th Annual Meeting of the Association for Computational Linguistics, pp. 5642–5650 (2020). https://aclanthology.org/2020.acl-main.499
19. Sennrich, R., Haddow, B., Birch, A.: Improving neural machine translation models with monolingual data. In: Proceedings of the 54th Annual Meeting of the Association for Computational Linguistics (Volume 1: Long Papers), pp. 86–96 (2016). https://aclanthology.org/P16-1009
20. Edunov, S., Ott, M., Auli, M., Grangier, D.: Understanding back-translation at scale. In: Proceedings of the 2018 Conference on Empirical Methods in Natural Language Processing, pp. 489–500 (2018). https://aclanthology.org/D18-1045
21. Singh, J., McCann, B., Keskar, N., Xiong, C., Socher, R.: XLDA: cross-lingual data augmentation for natural language inference and question answering. CoRR, **abs/1905.11471** (2019). http://arxiv.org/abs/1905.11471
22. Hou, Y., Liu, Y., Che, W., Liu, T.: Sequence-to-sequence data augmentation for dialogue language understanding. In: Proceedings of the 27th International Conference on Computational Linguistics, pp. 1234–1245 (2018). https://aclanthology.org/C18-1105
23. Kobayashi, S.: Contextual augmentation: data augmentation by words with paradigmatic relations. In: Proceedings of the 2018 Conference of the North American Chapter of the Association for Computational Linguistics: Human Language Technologies, Volume 2 (Short Papers), pp. 452–457 (2018). https://aclanthology.org/N18-2072
24. Kurata, G., Xiang, B., Zhou, B.: Labeled data generation with encoder-decoder LSTM for semantic slot filling. In: Proceedings of Interspeech 2016, pp. 725–729 (2016)
25. Chen, J., Yang, Z., Yang, D.: MixText: linguistically-informed interpolation of hidden space for semi-supervised text classification. In: Proceedings of the 58th Annual Meeting of the Association for Computational Linguistics, pp. 2147–2157 (2020). https://aclanthology.org/2020.acl-main.194
26. Kim, J., Choo, W., Song, H.: Puzzle Mix: exploiting saliency and local statistics for optimal mixup. In: Proceedings of the 37th International Conference on Machine Learning (2020)
27. Chen, H., Han, W., Yang, D., Poria, S.: DoubleMix: simple interpolation-based data augmentation for text classification. In: Proceedings of the 29th International Conference on Computational Linguistics, pp. 4622–4632 (2022). https://aclanthology.org/2022.coling-1.409
28. Xie, Q., Dai, Z., Hovy, E., Luong, T., Le, Q.: Unsupervised data augmentation for consistency training. Adv. Neural Inf. Process. Syst. **33**, 6256–6268 (2020). https://proceedings.neurips.cc/paper/2020/file/44feb0096faa8326192570788b38c1d1-Paper.pdf

29. Longpre, S., Wang, Y., DuBois, C.: How effective is task-agnostic data augmentation for pretrained transformers? In: Findings of the Association for Computational Linguistics: EMNLP 2020, pp. 4401–4411 (2020). https://aclanthology.org/2020.findings-emnlp.394
30. Wang, Z.: Linguistic knowledge in data augmentation for natural language processing: An example on Chinese question matching. In: Proceedings of the 5th International Conference on Natural Language and Speech Processing (ICNLSP 2022), pp. 40–49 (2022). https://aclanthology.org/2022.icnlsp-1.5
31. Wang, Z.: Random text perturbations work, but not always. In: Proceedings of the 3rd Workshop on Evaluation and Comparison of NLP Systems, pp. 51–57 (2022). https://aclanthology.org/2022.eval4nlp-1.6
32. Miller, G.: WordNet: A lexical database for English. Commun. ACM. **38**, 39–41 (1995). https://doi.org/10.1145/219717.219748
33. Jurafsky, D., Martin, J.: Speech and Language Processing: An Introduction to Natural Language Processing, Computational Linguistics, and Speech Recognition. Prentice Hall PTR (2009)
34. Brants, T., Popat, A., Xu, P., Och, F., Dean, J.: Large language models in machine translation. In: Proceedings of the 2007 Joint Conference on Empirical Methods in Natural Language Processing and Computational Natural Language Learning (EMNLP-CoNLL), pp. 858–867 (2007). https://aclanthology.org/D07-1090
35. Liu, X., Chen, Q., Deng, C., Zeng, H., Chen, J., Li, D., Tang, B.: LCQMC: A large-scale chinese question matching corpus. In: Proceedings of the 27th International Conference on Computational Linguistics, pp. 1952–1962 (2018). https://aclanthology.org/C18-1166
36. Mikolov, T., Chen, K., Corrado, G., Dean, J.: Efficient estimation of word representations in vector space. In: 1st International Conference on Learning Representations, ICLR 2013, Scottsdale, Arizona, USA, May 2–4, 2013, Workshop Track Proceedings (2013). http://arxiv.org/abs/1301.3781
37. Kim, Y.: Convolutional neural networks for sentence classification. In: Proceedings of the 2014 Conference on Empirical Methods in Natural Language Processing (EMNLP), pp. 1746–1751 (2014). https://aclanthology.org/D14-1181
38. Cho, K., Merriënboer, B., Gulcehre, C., Bahdanau, D., Bougares, F., Schwenk, H., Bengio, Y.: Learning phrase representations using RNN encoder–decoder for statistical machine translation. In: Proceedings of the 2014 Conference on Empirical Methods in Natural Language Processing (EMNLP), pp. 1724–1734 (2014). https://aclanthology.org/D14-1179
39. Hochreiter, S., Schmidhuber, J.: Long short-term memory. Neural Comput. **9**, 1735–1780 (1997). https://doi.org/10.1162/neco.1997.9.8.1735
40. Vaswani, A., Shazeer, N., Parmar, N., Uszkoreit, J., Jones, L., Gomez, A., Kaiser, L., Polosukhin, I.: Attention is All you Need. Adv. Neural Inf. Process. Syst. **30**, (2017). https://proceedings.neurips.cc/paper/2017/file/3f5ee243547dee91fbd053c1c4a845aa-Paper.pdf
41. Kingma, D., Ba, J.: Adam: A method for stochastic optimization. In: 3rd International Conference on Learning Representations, ICLR 2015, San Diego, CA, USA, May 7–9, 2015, Conference Track Proceedings (2015). http://arxiv.org/abs/1412.6980
42. Xiao, D., Li, Y., Zhang, H., Sun, Y., Tian, H., Wu, H., Wang, H.: ERNIE-Gram: Pre-training with explicitly n-gram masked language modeling for natural language understanding. In: Proceedings of the 2021 Conference of the North American Chapter of the Association for Computational Linguistics: Human Language Technologies, pp. 1702–1715 (2021). https://aclanthology.org/2021.naacl-main.136
43. Liu, B.: Sentiment Analysis and Opinion Mining. Morgan & Claypool (2012)

Scaling Up Paraphrase Generation Datasets with Machine Translation and Semantic Similarity Filtering

Besher Alkurdi, Hasan Yunus Sarioglu, and Mehmet Fatih Amasyali

1 Introduction

Paraphrase generation has a wide range of applications, such as data augmentation [16], machine translation evaluation [29], chatbots [14], question answering [35], and semantic parsing [9]. In a recent work, [1] addressed the scarcity of large-scale paraphrase datasets, particularly in non-English languages, by introducing high-quality and expansive paraphrase datasets in Turkish. In this chapter, we continue this work and present further contributions. We used English–Turkish datasets and translated the sentences from English to Turkish. We then employed a transformer-based [31] model to calculate the semantic similarity for each pair in the resulting Turkish–Turkish datasets. Pairs with a score exceeding a threshold were considered paraphrases. The threshold was chosen based on human annotations collected by our team.

This chapter presents several contributions. We introduce the largest Turkish paraphrase datasets currently available, comprising of roughly 800,000 pairs. Additionally, we propose a novel approach to constructing a paraphrase dataset from a parallel corpus, which involves combining machine translation and filtering based on semantic similarity. We provide paraphrase generation models that are trained on the datasets introduced in this chapter and assess their performance using diverse benchmark metrics. We also offer a manually annotated dataset on semantic textual similarity that contains 500 pairs.

To evaluate the effectiveness of the models trained on our paraphrase datasets, we employed various metrics for the paraphrase generation task. Moreover, we used

B. Alkurdi (✉) · H. Y. Sarioglu · M. F. Amasyali
Department of Computer Engineering, Yildiz Technical University, Esenler, Istanbul, Türkiye
e-mail: besher.alkurdi@std.yildiz.edu.tr; yunus.sarioglu@std.yildiz.edu.tr; amasyali@yildiz.edu.tr

M. Abbas (ed.), *Practical Solutions for Diverse Real-World NLP Applications*, Signals and Communication Technology,
https://doi.org/10.1007/978-3-031-44260-5_2

the generated paraphrases as a method for data augmentation, which allowed us to conduct an extrinsic evaluation of our models.

We have made our datasets and fine-tuned models publicly available[1] and hope that our work will stimulate further research in this area. Our datasets can be used to benchmark paraphrase generation architectures and datasets in the future.

2 Related Work

Identifying texts that convey similar or equivalent meanings, commonly referred to as paraphrase identification, poses a formidable challenge. Prior studies have attempted various methods to create datasets of paraphrases.

Collecting paraphrases manually is a costly and impractical endeavor, especially when resources are limited. Therefore, researchers in this field have frequently resorted to crowdsourcing paraphrase datasets [8]. This approach offers a noteworthy advantage, as it enables the construction of a high-quality dataset with increased sentence diversity, while minimizing the likelihood of generating pairs with low semantic similarity.

To detect paraphrases within a corpus of texts, semantic-similarity-based mining can be utilized. This involves comparing each sentence with every other sentence in the corpus and assigning a similarity score. The sentence with the highest score is deemed a paraphrase. However, due to its quadratic runtime, this method is not scalable for large paraphrase datasets. A comparable approach was adopted in a previous study [21].

One approach to creating paraphrases is by utilizing machine translation, where a text is translated into a pivot language and then translated back into the source language [28, 32]. This method can also incorporate multiple pivot languages in a similar manner. However, automatic translation from the source to the pivot language and back can introduce noise, which can affect the reliability of the results.

Various automatic approaches have also been employed to identify paraphrases, such as using parallel movie subtitles [3], image captions of the same image [20], and texts that can be marked as paraphrases based on different conditions, including duplicate questions,[2] duplicate posts [17], and text rewritings [22].

Several studies have been conducted on the creation of Turkish paraphrase datasets. For instance, [15] conducted a study resulting in 2472 text pairs that were annotated by humans. Additionally, [12] presented a paraphrase dataset consisting of 1270 paraphrase pairs from various sources. However, these datasets have not been shared publicly. More recently, [4] combined translated and manually generated datasets, focusing on question pairs, and trained a BERT2BERT architecture on

[1] https://github.com/mrbesher/semantic-filtering-for-paraphrasing.

[2] https://www.kaggle.com/c/quora-question-pairs.

it. Nevertheless, to the best of our knowledge, none of the existing studies provide a comprehensive paraphrase dataset in Turkish.

3 Dataset Creation

To ensure the high quality of the paraphrase dataset, we implemented a pipeline consisting of several steps. First, we downloaded English–Turkish parallel texts using Opus Tools. We considered several datasets shared on OPUS, including OpenSubtitles2018, TED2013, and Tatoeba v2022-03-03. We then pre-processed the text pairs according to the characteristics of each dataset, such as removing explanations done in the TED dataset indicated by two hyphens before and after the explanation.

Next, we applied machine translation to the entire dataset from English to Turkish. Afterward, we removed source and translated sentences that included each other and pre-processed the text pairs again to remove noisy texts generated by the translation model. Finally, we measured the semantic similarity between text pairs and selected pairs with high similarity scores as paraphrases. The steps of the pipeline are illustrated in Fig. 1. Through this robust process, we ensure that the dataset created is of high quality.

3.1 *Translation and Semantic-Similarity-Based Filtering*

As a result of the vast amount of data that required translation, it was impractical to utilize online machine translation services due to provider limitations. Instead,

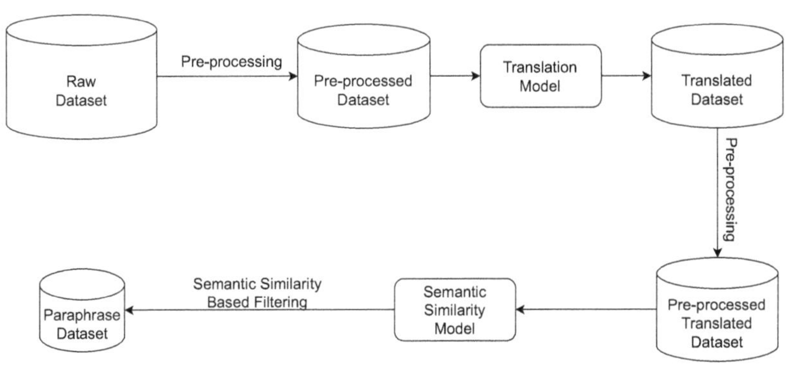

Fig. 1 Dataset creation pipeline

Table 1 The distribution of human annotations across the datasets

Label	OST	TAT	TED
No relevance	25	2	6
Distant meanings	43	15	26
Near synonyms	92	40	37
Synonyms	74	90	26

we opted for a machine translation model made available through the OPUS-MT project [30] and publicly accessible via Hugging Face.[3]

To further refine the pairs, we used a semantic similarity metric to exclude pairs with insufficient semantic similarity. Multiple models were available for semantic similarity scoring, and we needed to select the most appropriate one. To do so, we randomly selected 250, 150, and 100 pairs from OpenSubtitles2018, Tatoeba, and TED2013, respectively. These samples were then evaluated by six native Turkish speakers, with each pair assigned to two different annotators. As suggested in [11], each pair was assigned one of the labels indicating semantic similarity. In cases where the annotators disagreed and the score difference was less than two, we chose the label indicating lower semantic similarity. Otherwise, the label was discarded.

To simplify the process of collecting annotations, we developed a Telegram bot.[4] We utilized the scores provided by the annotators to establish a threshold for filtering out paraphrases of subpar quality.

During the annotation process, there were 16 samples from OpenSubtitles2018 (OST), 5 samples from TED2013 (TED), and 3 samples from Tatoeba (TAT) where the annotators differed by two points. Consequently, we discarded a total of 24 samples. The distribution of labels within each dataset is presented in Table 1.

In order to obtain high-quality paraphrase pairs, we focused on retaining phrase pairs labeled as near synonyms or synonyms in our analysis. Our examination of the TED, OST, and TAT datasets revealed that 66.32%, 70.94%, and 88.44% of the pairs, respectively, could be classified as paraphrases. As phrases with divergent meanings may negatively impact model performance, refining the filtering process was essential.

To determine the most suitable semantic similarity model, we tested various models on the manually labeled sample, evaluating their capacity to preserve valid pairs (i.e., those labeled as near synonyms or synonyms). We established model-specific thresholds to ensure that, after filtering out pairs with scores below these thresholds, 95% of the retained pairs were valid according to the human-annotated sample (Table 2).

Based on these criteria, we selected the semantic similarity model that maintained the highest percentage of valid pairs in the human-annotated sample, while minimizing the number of discarded pairs labeled as synonyms or near

[3] https://huggingface.co/Helsinki-NLP/opus-tatoeba-en-tr.

[4] https://telegram.org/.

Table 2 The number of text pairs in the datasets before and after filtering

Name	Raw	Pre-processing	Similarity-based filtering
OST	13,190,557	1,944,955	706,468
TAT	393,876	265,203	50,423
TED	131,874	104,238	39,763

synonyms. The chosen model,[5] fine-tuned on a machine-translated version of STS-b[6] and NLI [7], mapped texts to a 768-dimensional dense vector space. This model outperformed others, such as distiluse-base-multilingual-cased [25], and multilingual-112, which maps texts to a 384-dimensional dense vector space and is available on Huggingface.[7]

4 Evaluation

To assess the quality of our constructed datasets and establish a baseline for future research on Turkish paraphrase generation, we conducted a series of experiments. We trained our models on both the unfiltered and filtered versions of our datasets in order to analyze the effects of our applied filtering method on the quality of the datasets. One of the main uses of paraphrase generation models is in the area of data augmentation [13, 16]. To evaluate the effect of paraphrase generation using our models, we conducted some experiments and compared the effect of adding paraphrases to datasets concerning other tasks.

4.1 Paraphrase Generation

In order to conduct our experiments, a development split and test split were randomly selected from each dataset, with each split consisting of 5% of the pairs. The remaining pairs were used to train the models. This section presents the experimental results of our fine-tuned models, which were trained on the train splits and tested on the test splits of our datasets. Our approach employed transfer learning, utilizing pre-trained text-to-text transformer models such as mT5, which is a multilingual variant of T5 presented in [33]. To accomplish this, we made use of a pre-trained checkpoint of mT5-base provided by Google and published on Hugging Face.[8] Additionally, we employed BART [18], using a checkpoint of BART-base

[5] https://huggingface.co/emrecan/bert-base-turkish-cased-mean-nli-stsb-tr.

[6] https://huggingface.co/datasets/emrecan/stsb-mt-turkish.

[7] https://huggingface.co/sentence-transformers/paraphrase-multilingual-MiniLM-L12-v2.

[8] https://huggingface.co/google/mt5-base.

(uncased) that was pre-trained from scratch by [26]. The authors have made this model available on Hugging Face.[9]

Our initial experimentation with the TED dataset yielded unsatisfactory results, as the fine-tuned models were unable to produce acceptable paraphrases. Consequently, we chose not to proceed with further experimentation using this dataset. As a result, we are only able to present the translations and the filtered dataset, as we do not have any experimental results to report.

To train our models on the OST and TAT datasets, we used a learning rate of $1e-4$, with the models trained for 4 and 6 epochs, respectively. After conducting multiple experiments with varying learning rates, we found that these values resulted in the highest BLEU scores for the models on the development splits. For each source text, our models generated five candidate texts, and the candidate with the highest probability was selected for evaluation. Notably, in order to ensure diversity in the generated texts, we only chose candidates that did not consist of the same letters as the source.

Our report includes an analysis of several metrics, including BERTScore [34],[10] BLEU [23],[11] ROUGE [19], METEOR [5], and TER [27]. We present the mean of four results from four training runs, utilizing the settings previously described in this chapter. The scores can be found in Tables 3 and 4.

It is noteworthy that, in comparison to other models, the mT5-base trained on the OST dataset exhibited superior performance in both datasets. This finding suggests a high level of generalizability and dataset quality. To evaluate the effect of our filtering approach in greater detail, we fine-tuned the mT5-base on the unfiltered datasets. Results showed that although the unfiltered datasets were larger in size, the fine-tuned models demonstrated poorer performance on the OST dataset and generated fewer semantically similar pairs on the TAT dataset. Our belief is that the reason for this outcome is the more careful construction of TAT dataset, which involves crowdsourcing. As a result, the impact of semantic-similarity-based filtering is less apparent. In terms of results, we present the performance score of mT5-base trained for 3 and 4 epochs on the unfiltered OpenSubtitles2018 (OST-RAW) and Tatoeba (TAT-RAW), respectively. It is worth noting that the model's performance on the test sets began to decline after these epochs.

Tables 5 and 6 showcase a selection of paraphrases generated by the models that were fine-tuned on our train datasets. Each model's corresponding dataset abbreviation is indicated in parentheses. We selected examples that demonstrated both successful and unsuccessful paraphrasing examples.

[9] https://huggingface.co/mukayese/transformer-turkish-summarization.

[10] https://github.com/Tiiiger/bert_score.

[11] https://huggingface.co/spaces/evaluate-metric/bleu.

Table 3 The performance scores of our models on the test datasets. TER score measures distance. The other metrics measure similarity

Model	Train dataset	BERTScore cased	BERTScore uncased	BLEU	ROUGE-L	METEOR	TER
OST test dataset							
mT5-base	OST	**89 ± 0.01**	**92.05± 0.01**	**46.26± 0.09**	**74.8± 0.02**	**72.97 ± 0.13**	**36.4 ± 0.04**
trBART	OST	77.8 ± 0.17	87.92 ± 0.13	33.59 ± 0.32	64.65 ± 0.33	62.62 ± 0.45	50.96 ± 0.4
mT5-base	TAT	84.95 ± 0.38	89.23 ± 0.24	29.37 ± 0.83	66.64 ± 0.46	63.14 ± 0.86	49.29 ± 1.13
trBART	TAT	74.21 ± 0.32	85.25 ± 0.28	23.45 ± 0.29	59.32 ± 0.6	54.93 ± 0.5	57.22 ± 0.59
TAT test dataset							
mT5-base	TAT	94.07 ± 0.36	95.75 ± 0.25	61.66 ± 1.34	84.67 ± 0.62	**82.72 ± 0.42**	22.43 ± 1.27
trBART	TAT	84.42 ± 0.33	94.09 ± 0.26	56.58 ± 0.99	81.68 ± 0.54	78.83 ± 0.52	26.69 ± 0.76
mT5-base	OST	**94.47 ± 0.06**	**95.94 ± 0.03**	**63.87 ± 0.44**	**85.18 ± 0.19**	82.46 ± 0.27	**21.41 ± 0.21**
trBART	OST	82.65 ± 0.25	92.47 ± 0.16	48.71 ± 0.69	76.45 ± 0.32	73.26 ± 0.6	34.79 ± 0.28

Table 4 A Comparison Between the Performance of mT5 Model Checkpoints Trained on Our Filtered and Unfiltered Datasets. TER score measures distance. The other metrics measure similarity

Model	Train dataset	BERTScore cased	BERTScore uncased	BLEU	ROUGE-L	METEOR	TER
OST test dataset							
mT5-base	OST	**89 ± 0.01**	**92.05 ± 0.01**	**46.26 ± 0.09**	**74.8 ± 0.02**	**72.97 ± 0.13**	**36.4 ± 0.04**
mT5-base	OST *(Unfiltered)*	88.89 ± 0.06	91.94 ± 0.04	36.4 ± 0.23	73.87 ± 0.09	72.16 ± 0.16	37.58 ± 0.15
mT5-base	TAT	84.95 ± 0.38	89.23 ± 0.24	29.37 ± 0.83	66.64 ± 0.46	63.14 ± 0.86	49.29 ± 1.13
mT5-base	TAT *(Unfiltered)*	88.95 ± 0.2	92.08 ± 0.14	38.13 ± 0.4	68.39 ± 0.23	65.87 ± 0.22	45.13 ± 0.33
TAT test dataset							
mT5-base	TAT	94.07 ± 0.36	95.75 ± 0.25	61.66 ± 1.34	84.67 ± 0.62	82.72 ± 0.42	22.43 ± 1.27
mT5-base	TAT *(Unfiltered)*	91.61 ± 0.12	93.93 ± 0.09	34.74 ± 0.62	**86.6 ± 0.2**	**84.85 ± 0.23**	**18.23 ± 0.25**
mT5-base	OST	**94.47 ± 0.06**	**95.94 ± 0.03**	**63.87 ± 0.44**	85.18 ± 0.19	82.46 ± 0.27	21.41 ± 0.21
mT5-base	OST *(Unfiltered)*	91.97 ± 0.07	94.2 ± 0.05	37.02 ± 0.16	84.05 ± 0.19	81.59 ± 0.28	22.76 ± 0.32

Table 5 Generated paraphrases of examples from the OST dataset

Source	Ve onu sizden kimse alamaz, beyler	Woodhouse tatlım biraz daha buza ihtiyacım var.	Bir sandviçe yetecek kadar malzemem var
	And no one can take it away from you, gentlemen	*Woodhouse, honey, I need some more ice*	*I've got stuff that will be enough for a sandwich*
mT5-base (OST)	Ve kimse onu sizden alamaz, beyler	Woodhouse, tatlım, biraz daha buz lazım	Bir sandviç için yeterli malzemem var
	And no one can take it away from you, gentlemen	*Woodhouse, honey, there is a need for more ice*	*I've got enough stuff for a sandwich*
trBART (OST)	beyler ve onu sizden kimse alamaz	woodhouse biraz daha buza ihtiyacım var	bir sandviçe yetecek kadar malzeme var
	Gentlemen and no one can take it away from you	*woodhouse, I need some more ice*	*There is enough stuff for a sandwich*
mT5-base (TAT)	Kimse bunu sizden alamaz, beyler	Woodhouse tatlım biraz daha buza ihtiyacın var	Sandviçe yetecek kadar malzemem var
	No one can take that away from you, gentlemen	*Woodhouse, honey, you need more ice*	*I've got stuff that will be enough for sandwich*
trBART (TAT)	beyler ve onu sizden kimse alamaz	woodhouse biraz daha buza ihtiyacım var	bir sandviçe yetecek kadar malzeme var
	gentlemen and no one can take it away from you	*woodhouse, honey, there is a need for more ice*	*there is enough stuff for a sandwich*
mT5-base (OST-RAW)	Kimse onu senden alamaz, çocuklar	Woodhouse, tatlım, biraz daha buz lazım	Bir sandviç için yeterli malzemem var
	No one can take it away from you, kids	*Woodhouse, honey, there is a need for more ice*	*I've got enough stuff for a sandwich*
mT5-base (TAT-RAW)	Kimse bunu sizden alamaz, beyler	Woodhouse tatlım biraz daha buza ihtiyacın var	Sandviçe yetecek kadar malzemem var
	No one can take that away from you, gentlemen	*Woodhouse honey you need more ice*	*I've got stuff that will be enough for sandwich*

Table 6 Generated paraphrases of examples from the TAT dataset

Source	Tom daha sonra ne yapacağını bilmiyordu	Tom asla tek başına oraya gitmezdi	İlk olarak ne yapacaklarını merak ettiler
	Tom didn't know what to do next	*Tom would never go there by himself*	*They wondered what they would do first*
mT5-base (OST)	Tom ne yapacağını bilmiyordu	Tom oraya tek başına gitmezdi	Önce ne yapacaklarını merak ettiler
	Tom didn't know what to do	*Tom wouldn't go there by himself*	*They wondered what they would do before*
trBART (OST)	tom bundan sonra ne yapacağını bilmiyordu	tom oraya hiç gitmezdi	ilk olarak ne yapacaklarını merak ediyorlar
	tom didn't know what to do next	*tom never went there*	*they are wondering what they're going to do first*
mT5-base (TAT)	Tom sonra ne yapacağını bilmiyordu	Tom oraya asla tek başına gitmez	İlk başta ne yapacaklarını merak ettiler
	Tom didn't know what to do next	*Tom never goes there by himself*	*They wondered what they were going to do at first*
trBART (TAT)	tom bundan sonra ne yapacağını bilmiyordu	tom oraya hiç gitmezdi	ilk olarak ne yapacaklarını merak ediyorlar
	tom didn't know what to do next	*tom never went there*	*they are wondering what they're going to do first*
mT5-base (OST-RAW)	Tom bundan sonra ne yapacağını bilmiyordu	Tom oraya hiç tek başına gitmedi	Önce ne yapacaklarını merak ediyorlar
	tom didn't know what to do next	*Tom didn't go there by himself*	*They are wondering what they would do before*
mT5-base (TAT-RAW)	Tom bundan sonra ne yapacağını bilmiyordu	Tom oraya asla tek başına gitmez	Önce ne yapacaklarını merak ettiler
	Tom didn't know what to do next	*Tom never goes there by himself*	*They wondered what they would do before*

4.2 Data Augmentation

In natural language processing, the availability of large amounts of data is crucial for achieving high performance. Collecting and annotating large datasets is a costly and time-consuming process; for this reason, several data augmentation techniques have been introduced to address this issue. We used data augmentation as an indirect task to evaluate our paraphrase generation models. If the generated paraphrases are similar to the original texts, the performance will not be affected. If the paraphrases are of low quality and have distant meanings, the performance is expected to decrease. In order to carry out our assessment, we conducted experiments on two Turkish text classification datasets: Offenseval [10] and Turkcell-17k [2].[12] The OffensEval dataset is an offensive language detection dataset and consists of 28,000 training samples. The Turkcell-17k dataset is a sentiment analysis dataset that we partitioned into train and test sets and comprises 13,832 training samples. We also utilized the Semantic Textual Similarity Benchmark (STS-B) dataset, which was shared in [6] and has about 5750 training samples.

We used a distilled version of the BERT model[13] for the text classification task, employing Adam as the optimization algorithm with a batch size of 32 and a learning rate of $1e-5$. Our experimentation involved training models with and without augmentation, using smaller subsets of the datasets. The models trained on augmented datasets were trained for half the number of epochs as the models trained on original datasets. For the STS-B dataset, we fine-tuned a pre-trained BERT model[14] in line with [24]. Throughout the training process, we recorded the highest scores achieved for each task. Macro-averaged F1 and accuracy were computed for text classification, while Pearson correlation coefficient was used for semantic textual similarity.

Tables 7 and 8 present the findings of our experiments, and Figs. 2, 3, and 4 provide visual representations of the sample ratio vs. performance. The experimental results demonstrate that the performance of the models increases with the size of the dataset. However, our results indicate that data augmentation, particularly through paraphrasing, can significantly improve model performance, especially when the dataset is small. We observe that the performance gap between the augmented and original datasets narrows as a larger subset of the dataset is used. Our findings suggest that our paraphrasing models have a positive impact on the results especially in scenarios where only a limited amount of data is available.

[12] http://www.kemik.yildiz.edu.tr/veri_kumelerimiz.html.

[13] https://huggingface.co/dbmdz/distilbert-base-turkish-cased.

[14] https://huggingface.co/dbmdz/bert-base-turkish-cased.

Table 7 Comparing the impact of paraphrase augmentation on classification performance using distilled Turkish BERT model

Dataset	Sample ratio	F1 macro avg.		Accuracy	
		Original	Augmented	Original	Augmented
OffensEval	5%	58.69%	**69.23%**	80.88%	**82.67%**
	10%	71.06%	**74.40%**	83.81%	**85.40%**
	25%	74.67%	**75.95%**	85.63%	**86.03%**
	50%	77.36%	**77.80%**	86.77%	**86.91%**
	75%	77.93%	**78.57%**	87.14%	**87.60%**
	100%	**79.05%**	79.02%	**87.76%**	87.48%
Turkcell-17k	5%	45.37%	**52.13%**	50.07%	**55.19%**
	10%	52.61%	**61.71%**	55.53%	**62.68%**
	25%	64.64%	**67.62%**	65.69%	**68.38%**
	50%	66.71%	**70.62%**	67.46%	**71.42%**
	75%	70.44%	**72.03%**	71.30%	**72.75%**
	100%	72.09%	**73.92%**	72.66%	**74.37%**

Table 8 Comparing the impact of paraphrase augmentation on semantic textual similarity performance using Turkish BERT model

Sample ratio	Pearson correlation	
	Original	Augmented
5%	0.8030	**0.8061**
10%	0.8024	**0.8066**
25%	0.8053	**0.8096**
50%	0.8088	**0.8094**
75%	**0.8137**	0.8105
100%	**0.8165**	0.8143

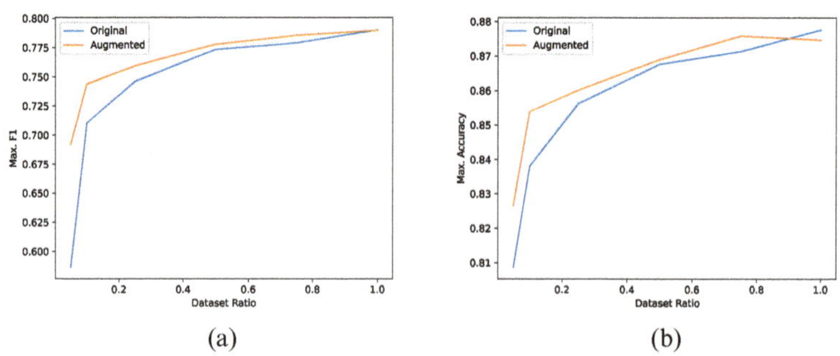

Fig. 2 The impact of the dataset ratio on accuracy and F1 for the OffensEval dataset. (**a**) Macro-averaged F1 score. (**b**) Accuracy

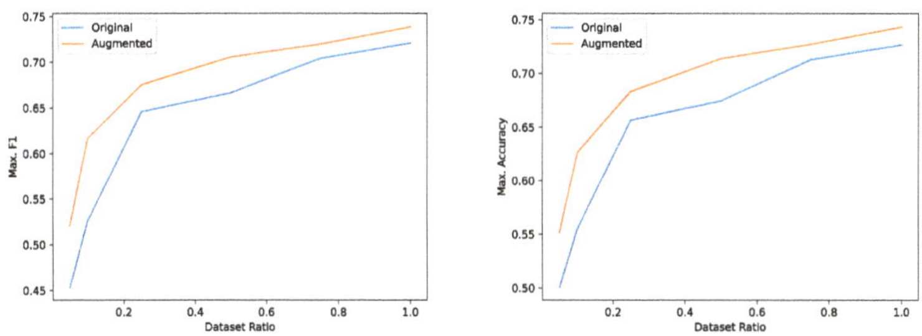

(a) (b)

Fig. 3 The impact of the dataset ratio on accuracy and F1 for the Turkcell-17k dataset. (**a**) Macro-averaged F1 score. (**b**) Accuracy

Fig. 4 The impact of the dataset ratio on pearson correlation for the STS benchmark dataset

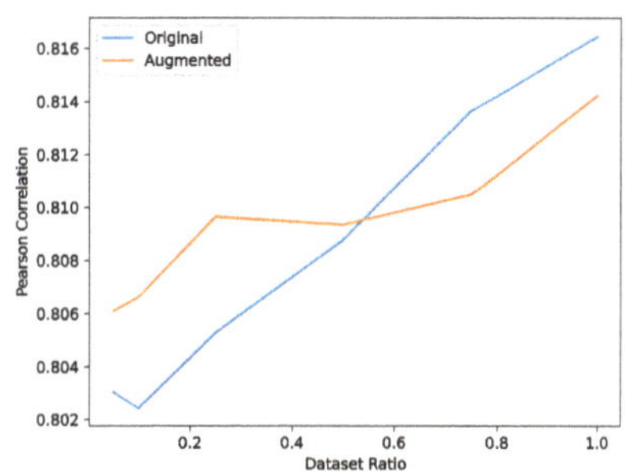

5 Conclusion

In this chapter, we presented an approach for constructing paraphrase datasets from parallel text corpora, which leverages machine translation and filtering based on semantic similarity. Our methodology involved selecting a semantic similarity model that retained the most paraphrases in the datasets, according to similarity ratings obtained from human annotators. We showcased the effectiveness of our approach by presenting the resulting paraphrase datasets and reporting benchmarking results of text-to-text transformer models trained on these datasets. Additionally, we evaluated the trained models by using them for data augmentation, demonstrating their effectiveness in real-world use cases. Our findings suggest that the proposed methodology can significantly contribute to the development of improved paraphrase generation datasets.

Our contributions can be summarized as follows:

- We proposed a novel approach for constructing paraphrase datasets from parallel corpora, combining machine translation and filtering based on semantic similarity.
- We introduced the largest Turkish paraphrase datasets currently available, comprising approximately 800,000 pairs.

- We provided paraphrase generation models that are trained on the datasets introduced in this chapter and assessed their performance using various benchmark metrics.
- We offered a manually annotated dataset on semantic textual similarity that contains 500 pairs.
- We conducted data augmentation experiments using the trained paraphrase generation models, allowing for an extrinsic evaluation of our models.

Acknowledgments This study was supported by the Scientific and Technological Research Council of Turkey (TUBITAK) Grant No: 120E100. The authors declare no conflicts of interest, financial or otherwise, with the aforementioned organization.

References

1. Alkurdi, B., Sarioglu, H.Y., Amasyali, M.F.: Semantic similarity based filtering for Turkish paraphrase dataset creation. In: Proceedings of the 5th International Conference on Natural Language and Speech Processing (ICNLSP 2022), pp. 119–127, Trento, Italy, December 2022. Association for Computational Linguistics
2. Amasyali, M.F., Tasköprü, H., Çaliskan, K.: Words, meanings, characters in sentiment analysis. In: 2018 Innovations in Intelligent Systems and Applications Conference (ASYU), pp. 1–6, Oct 2018
3. Aulamo, M., Sulubacak, U., Virpioja, S., Tiedemann, J.: OpusTools and parallel corpus diagnostics. In: Proceedings of the 12th Language Resources and Evaluation Conference, pp. 3782–3789. European Language Resources Association, May 2020
4. Bağcı, A., Amasyali, M.F.: Comparison of Turkish paraphrase generation models. In: 2021 International Conference on INnovations in Intelligent SysTems and Applications (INISTA), pp. 1–6. IEEE (2021)
5. Banerjee, S., Lavie, A.: METEOR: An automatic metric for MT evaluation with improved correlation with human judgments. In: Proceedings of the ACL Workshop on Intrinsic and Extrinsic Evaluation Measures for Machine Translation and/or Summarization, pp. 65–72, Ann Arbor, Michigan, June 2005. Association for Computational Linguistics
6. Beken Fikri, F., Oflazer, K., Yanikoglu, B.: Semantic similarity based evaluation for abstractive news summarization. In: Proceedings of the 1st Workshop on Natural Language Generation, Evaluation, and Metrics (GEM 2021), pp. 24–33, Online, August 2021. Association for Computational Linguistics
7. Budur, E., Özçelik, R., Gungor, T., Potts, C.: Data and representation for Turkish natural language inference. In: Proceedings of the 2020 Conference on Empirical Methods in Natural Language Processing (EMNLP), pp. 8253–8267, Online, November 2020. Association for Computational Linguistics
8. Burrows, S., Potthast, M., Stein, B.: Paraphrase acquisition via crowdsourcing and machine learning. ACM Trans. Intell. Syst. Technol. (TIST) **4**(3), 1–21 (2013)
9. Cao, R., Zhu, S., Yang, C., Liu, C., Ma, R., Zhao, Y., Chen, L., Yu, K.: Unsupervised dual paraphrasing for two-stage semantic parsing. Preprint (2020). arXiv:2005.13485
10. Çöltekin, Ç.: A corpus of Turkish offensive language on social media. In: Proceedings of the 12th Language Resources and Evaluation Conference, pp. 6174–6184, Marseille, France, 2020
11. Creutz, M.: Open subtitles paraphrase corpus for six languages. In: Proceedings of the Eleventh International Conference on Language Resources and Evaluation (LREC 2018), Miyazaki, Japan, May 2018. European Language Resources Association (ELRA)

12. Demir, S., El-Kahlout, I.D., Unal, E.: A case study towards Turkish paraphrase alignment. In: Proceedings of the 14th European Workshop on Natural Language Generation, pp. 188–192, 2013
13. Gao, S., Zhang, Y., Ou, Z., Yu, Z.: Paraphrase augmented task-oriented dialog generation. In: Proceedings of the 58th Annual Meeting of the Association for Computational Linguistics, pp. 639–649, Online, July 2020. Association for Computational Linguistics
14. Garg, S., Prabhu, S., Misra, H., Srinivasaraghavan, G.: Unsupervised contextual paraphrase generation using lexical control and reinforcement learning. Preprint (2021). arXiv:2103.12777
15. Karaoğlan, B., Kışla, T., Metin, S.K., Hürriyetoğlu, U., Soleymanzadeh, K.: Using multiple metrics in automatically building Turkish paraphrase corpus. Res. Comput. Sci. **117**, 75–83 (2016)
16. Kumar, A., Bhattamishra, S., Bhandari, M., Talukdar, P.: Submodular optimization-based diverse paraphrasing and its effectiveness in data augmentation. In: Proceedings of the 2019 Conference of the North American Chapter of the Association for Computational Linguistics: Human Language Technologies, Volume 1 (Long and Short Papers), pp. 3609–3619, 2019
17. Lan, W., Qiu, S., He, H., Xu, W.: A continuously growing dataset of sentential paraphrases. In: Proceedings of the 2017 Conference on Empirical Methods in Natural Language Processing, pp. 1224–1234, Copenhagen, Denmark, September 2017. Association for Computational Linguistics
18. Lewis, M., Liu, Y., Goyal, N., Ghazvininejad, M., Mohamed, A., Levy, O., Stoyanov, V., Zettlemoyer, L.: BART: Denoising sequence-to-sequence pre-training for natural language generation, translation, and comprehension. In: Proceedings of the 58th Annual Meeting of the Association for Computational Linguistics, pp. 7871–7880, Online, July 2020. Association for Computational Linguistics
19. Lin, C.-Y.: ROUGE: A package for automatic evaluation of summaries. In: Text Summarization Branches out, pp. 74–81, Barcelona, Spain, July 2004. Association for Computational Linguistics
20. Lin, T.-Y., Maire, M., Belongie, S., Hays, J., Perona, P., Ramanan, D., Dollár, P., Zitnick, C.L.: Microsoft COCO: Common objects in context. In: European Conference on Computer Vision, pp. 740–755. Springer (2014)
21. Martin, L., Fan, A., de la Clergerie, É., Bordes, A., Sagot, B.: MUSS: multilingual unsupervised sentence simplification by mining paraphrases. Preprint (2020). arXiv:2005.00352
22. Max, A., Wisniewski, G.: Mining naturally-occurring corrections and paraphrases from Wikipedia's revision history. In: Proceedings of the Seventh International Conference on Language Resources and Evaluation (LREC'10), Valletta, Malta, May 2010. European Language Resources Association (ELRA)
23. Papineni, K., Roukos, S., Ward, T., Zhu, W.-J.: BLEU: a method for automatic evaluation of machine translation. In: Proceedings of the 40th Annual Meeting of the Association for Computational Linguistics, pp. 311–318, 2002
24. Reimers, N., Gurevych, I.: Sentence-BERT: Sentence embeddings using Siamese Bert-networks. In: Proceedings of the 2019 Conference on Empirical Methods in Natural Language Processing. Association for Computational Linguistics, 11 2019
25. Reimers, N., Gurevych, I.: Making monolingual sentence embeddings multilingual using knowledge distillation. In: Proceedings of the 2020 Conference on Empirical Methods in Natural Language Processing (EMNLP), pp. 4512–4525, Online, November 2020. Association for Computational Linguistics
26. Safaya, A., Kurtuluş, E., Goktogan, A., Yuret, D.: Mukayese: Turkish NLP strikes back. In: Findings of the Association for Computational Linguistics: ACL 2022, pp. 846–863, Dublin, Ireland, May 2022. Association for Computational Linguistics
27. Snover, M., Dorr, B., Schwartz, R., Micciulla, L., Makhoul, J.: A study of translation edit rate with targeted human annotation. In: Proceedings of the 7th Conference of the Association for Machine Translation in the Americas: Technical Papers, pp. 223–231, Cambridge, Massachusetts, USA, 8 2006. Association for Machine Translation in the Americas

28. Suzuki, Y., Kajiwara, T., Komachi, M.: Building a non-trivial paraphrase corpus using multiple machine translation systems. In: Proceedings of ACL 2017, Student Research Workshop, pp. 36–42, Vancouver, Canada, July 2017. Association for Computational Linguistics
29. Thompson, B., Post, M.: Automatic machine translation evaluation in many languages via zero-shot paraphrasing. In: Proceedings of the 2020 Conference on Empirical Methods in Natural Language Processing (EMNLP), pp. 90–121, Online, November 2020. Association for Computational Linguistics
30. Tiedemann, J., Thottingal, S.: OPUS-MT – Building open translation services for the World. In: Proceedings of the 22nd Annual Conference of the European Association for Machine Translation (EAMT), Lisbon, Portugal, 2020
31. Vaswani, A., Shazeer, N., Parmar, N., Uszkoreit, J., Jones, L., Gomez, A.N., Kaiser, L., Polosukhin, I.: Attention is all you need. CoRR, abs/1706.03762, 2017
32. Wieting, J., Gimpel, K.: ParaNMT-50M: Pushing the limits of paraphrastic sentence embeddings with millions of machine translations. In: Proceedings of the 56th Annual Meeting of the Association for Computational Linguistics (Volume 1: Long Papers), pp. 451–462, Melbourne, Australia, July 2018. Association for Computational Linguistics
33. Xue, L., Constant, N., Roberts, A., Kale, M., Al-Rfou, R., Siddhant, A., Barua, A., Raffel, C.: mT5: A massively multilingual pre-trained text-to-text transformer. In: Proceedings of the 2021 Conference of the North American Chapter of the Association for Computational Linguistics: Human Language Technologies, pp. 483–498, Online, June 2021. Association for Computational Linguistics
34. Zhang, T., Kishore, V., Wu, F., Weinberger, K.Q., Artzi, Y.: BERTScore: Evaluating text generation with BERT. Preprint (2019). arXiv:1904.09675
35. Zhu, S., Cheng, X., Su, S., Lang, S.: Knowledge-based question answering by jointly generating, copying and paraphrasing. In: Proceedings of the 2017 ACM on Conference on Information and Knowledge Management, pp. 2439–2442, 2017

Generative Byte-Level Models for Restoring Spaces, Punctuation, and Capitalization in Multiple Languages

Laurence Dyer, Anthony Hughes, and Burcu Can

1 Introduction

Correct spacing, capitalization, and punctuation are all vital elements that contribute both to the readability of texts by humans and to their efficacy in a variety of natural language processing (NLP) applications [14, 19, 21]. Techniques for accurately restoring these features are therefore necessary when dealing with text data from which they are absent. Much work to date has focused on restoration of capitalization and/or punctuation—motivated by its application in processing the outputs of ASR systems [9]—but we include spaces as a restoration target, thereby extending the range of possible use cases to include parsing of hashtags, URLs, and other types of noisy or unprocessed text.

The omission of spaces from input texts introduces a new aspect to the restoration problem. Traditional token-based approaches are no longer sufficient because tokens are not known at input time. Processing of text at the character level or below is therefore required.

In a previous work on this problem [8], a range of pipeline models that combined character- and token-based approaches were compared with an end-to-end character-based model that restored all features under consideration in a

†Laurence Dyer and Anthony Hughes contributed equally to this work as first authors.

L. Dyer (✉) · A. Hughes
University of Wolverhampton, Wolverhampton, UK
e-mail: L.J.Dyer@wlv.ac.uk; A.J.Hughes2@wlv.ac.uk

B. Can
University of Stirling, Stirling, UK
e-mail: Burcu.Can@stir.ac.uk

M. Abbas (ed.), *Practical Solutions for Diverse Real-World NLP Applications*, Signals and Communication Technology,
https://doi.org/10.1007/978-3-031-44260-5_3

single inference step. The authors demonstrated the feasibility of the end-to-end approach by presenting a model that outperformed pipeline models proposed in existing literature and noted other advantages to the character-based approach, such as the possibility of restoring punctuation in *scriptio continua* languages directly without the need for a word segmentation step[1] and ability to restore mid-token capitalization and punctuation such as that found in "iPhone" or "yahoo.com." However, a pipeline including a fine-tuned token-level transformer layer was found to outperform the character-based model overall.

In this work, we combine the token-free approach explored in [8] with the power of pre-trained transformers by investigating the applicability of byte-level pre-trained transformers to the restoration of features for three languages—English, Japanese, and Gujarati—from distinct language families and with very distinct linguistic properties (which are detailed in Sect. 4.1).

2 Related Work

In this section, we present an overview of existing work on restoration of spaces, and of punctuation and capitalization, as separate tasks, and where recent work has combined these tasks using end-to-end and/or pipeline models, and finally of applications of byte-level transformer models to other NLP tasks.

2.1 Space Restoration/Word Segmentation

The task of restoring spaces in spaced languages is similar to that of word segmentation for *scriptio continua* languages, for which an extensive body of research exists due to its applications in downstream NLP tasks. Work in this area traditionally focused on the use of recurrent neural networks (RNNs) [16, 27, 39], but pre-trained languages models (PLMs) and transformers [6] have become the predominate architectures in recent years.

Attention is one of the key mechanisms employed in transformer models to overcome the challenges faced when using RNNs [33]. Reference [13] shows that employing character attention in a character-based neural network can produce state-of-the-art results for Japanese word segmentation. A similar approach presented in [3] builds upon this research by introducing attention for representation of

[1] In this work, we tackle restoration of spaces in the spaced languages English and Gujarati, as distinct from the task of word segmentation for *scriptio continua* languages, for which a wealth of literature exists. While the two tasks are related, the former refers to restoration of a feature that occurs naturally in texts produced in the language under consideration, whereas the latter refers to artificial segmentation of texts in *scriptio continua* languages.

multiple linguistic aspects such as word, subword, and character clusters to improve performance for Thai word segmentation.

Duan and Zhao [7] build upon earlier transformer research [6] by applying a custom Gaussian-masked directional transformer to the problem of Chinese word segmentation. Due to the limited testing of this model, Maimaiti et al. [18] build a more robust architecture that can generalize to more domains by using self-supervised learning to fine-tune an existing PLM, BERT. Further research [22] has attempted to tackle the issue of generalization using unsupervised methods and PLMs.

2.2 Restoration of Capitalization and Punctuation

Recent work has demonstrated the advantages of using transformer architectures for restoration of capitalization and punctuation [2, 11, 20, 32].

In [20], a single model using a transformer-enriched sequence-to-sequence LSTM model is proposed for the restoration of capitalization and punctuation. [34] proposed another encoder–decoder transformer architecture with attention for the restoration of punctuation only. Since these models operate at the token level, mixed-case words such as "iOS" are mapped to title case words. The authors [30] tackle this issue with character-level recurrent neural networks for truecasing only, demonstrating the efficacy of token-free architectures.

Multiple studies [5, 38] have shown that state-of-the-art results for punctuation restoration can be achieved using encoder-only models such as BERT [6] and RoBERTa [17]. The results in these works were improved upon in [10] by training and evaluating models at chunk level rather than segment level.

Models that are able to restore textual features to multilingual corpora have been shown to be highly successful. In [10], a single model for punctuation restoration is trained using XLM-RoBERTa [4], although it is observed that better results for English are obtained when using the monolingual RoBERTa. In [12], IndicBERT [15] is applied to a corpus of 11 Indic languages (which includes Gujarati), further demonstrating the efficacy of a single-model approach.

2.3 Restoration of Spaces, Capitalization, and Punctuation

As far as we are aware, only two studies have tackled the joint task of restoring spaces, punctuation, and capitalization. In the first of these, [28], the authors employ bigram-level gated recurrent units (GRUs) in a pipeline where each component restores one feature sequentially. In [8], a variety of pipeline and end-to-end approaches are compared on the same set of tasks. The best performing model overall for English was a pipeline consisting of a Naive Bayes-based model for space restoration and a fine-tuned transformer model for punctuation and capitalization.

However, a character-level vanilla BiLSTM model without pre-training was also found to be able to restore all features studied for English, Japanese, and Gujarati with some degree of accuracy.

2.4 Text Normalization/Diacritization with Byte-Level Transformers

Given the success of token-free NLP models and transformers, in this section we review the literature at the intersection of these two streams of research. As far as we are aware, no previous work has applied these tools to the restoration of spaces, capitalization, and punctuation, so we expand the scope of our review to include other NLP tasks.

Early work on character-level transformers [1] showed that they can outperform RNNs in the problem of language modelling. ByT5 [36], a variant of T5, improved on these initial works by using a much larger and multilingual training set called the "Colossal Cleaned Crawled Corpus."[2] The creators of ByT5 found that the best results were obtained when using an imbalanced encoder–decoder stack where the depth of the encoder was greater than that of the decoder.

A two-stage approach to training ByT5 has been shown to be effective for the problem of text normalization [26]. In the first step, the pre-training task is repeated using synthetically generated data relevant to text normalization. In the second step, the model is fine-tuned with authentic data. The authors note that their proposed solution is not entirely language-agnostic, as linguistic knowledge is required for the creation of the correct synthetic data for use in the first step. The authors in [25] present promising results for text normalization on code-mixed data, although that study is limited to a dataset consisting only of Indonesian and English.

A study on diacritization and typographical error correction found that combining ByT5 with classical dictionary methods can improve model accuracy for lesser-resourced languages such as Hungarian and Latvian [29]. However, the hybrid approach was found to be less effective for languages where more data were available.

3 Byte-Level Transformer Architecture

In this section, we outline the theoretical background to our work by describing encoder–decoder models, byte-level transformers, and the pre-trained ByT5 model that we employ in our experiments.

[2] https://www.tensorflow.org/datasets/catalog/c4.

3.1 Encoder–Decoder Overview

An encoder–decoder model is a popular machine learning architecture used for sequence-to-sequence tasks. In neural-based encoder–decoder models, each encoder and decoder is a neural network that operates on input and output sequences, respectively.

The encoder is the component responsible for processing the input sequence and creating a fixed-length vector called a context vector. In a byte-level text model, the input text is first converted to a sequence of bytes, which are then fed into the encoder network. At each step within an epoch, the network updates its hidden state based on the current input byte and the previous hidden state, until it reaches the end of the input sequence. The final hidden state of the encoder is then used to create the context vector, which the decoder can then utilize for learning a matching output.

The decoder is responsible for generating the output sequence based on the context vector produced by the encoder. In a byte-level text model, the output text is also represented as a sequence of bytes, and the decoder uses the context vector and the previous output byte to generate the next byte in the sequence. At each time step, the decoder takes the previous output byte, the context vector, and the previous hidden state as inputs and produces the next hidden state and output byte as outputs.

In order to train an encoder–decoder model, a form of parallel corpus is required. In the case of the models developed in this work, an input text such as `hellotherethisisasentence` would have a parallel text `Hello there, this is a sentence.`. When training the model, the goal is to minimize the difference between its predicted output and the true output. Given an input byte sequence $(x_1, \ldots, x_T)$ and a target sequence $(y_1, \ldots, y_{T'})$, a loss function is then used to measure the distance between the predicted and true probability distributions $p(y_1, \ldots, y_{T'} \mid x_1, \ldots, x_T)$ over all possible output bytes at each time step. Softmax is often employed for this purpose as it can calculate the distribution used over all the possible bytes in the input. A generic output function is defined as follows:

$$p(y_1, \ldots, y_{T'} \mid x_1, \ldots, x_T) = \prod_{t=1}^{T'} p(y_t \mid v, y_1, \ldots, y_{t-1}) \tag{1}$$

where v is a vector that captures the current state of the model. Further discussion of context vectors in transformer models follows in Sect. 3.2.

3.2 Byte-Level Transformers

The current literature experimenting with byte-level transformers [1, 36] closely follows [33]. The key difference between the original encoder–decoder model [31] and transformer models is the use of a stacked neural network with a self-attention

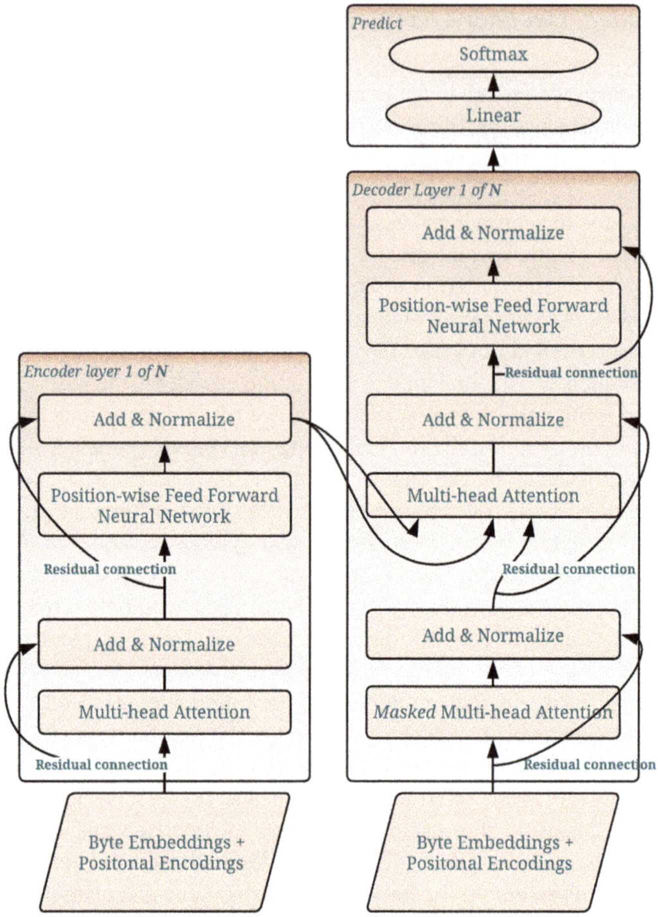

Fig. 1 Model architecture of a byte-level transformer

mechanism. Models from deep LSTM experiments are now being outperformed through the use of multi-head attention networks connected to feed-forward layers. A vanilla transformer model architecture for byte-level tasks is presented in Fig. 1.

3.2.1 Self-Attention

Self-attention is a mechanism that computes a weighted sum of values based on their relevance to a given query. This mechanism is an attention function which maps a query vector Q and a set of key–value (K–V) pairs to an output vector. The function computes a compatibility score between the query Q and each key K, which then gives a weighting to the corresponding value, as shown in Eq. 2 below. The weighted values are then aggregated into a single output vector. This process allows the self-attention mechanism to capture dependencies between different positions of an input sequence.

In byte-level transformer models, a decoder often employs masked self-attention to improve computational efficiency through parallelization, as shown in Eq. 3 below. The structure is similar to multi-head self-attention. However, the output of the softmax function is masked to prevent dependencies on future bytes. Specifically, the model generates predictions using only the left context of the byte being predicted. This results in the model attending solely to the preceding words in the sequence when generating the next byte based on that context.

The mathematical definition of multi-head attention is as follows:

$$\text{head}_i = \text{SelfAttention}(Q\mathbf{W}_i^Q, K\mathbf{W}_i^K, V\mathbf{W}_i^V) \tag{2}$$

$$\text{MultiHead}(Q, K, V) = \text{Concat}(\text{head}_1, \ldots, \text{head}_h)\mathbf{W}^O \tag{3}$$

where **W** refers to corresponding weight matrices.

3.3 *ByT5*

ByT5 is a multilingual token-free large language model (LLM). It is based on mT5 [37] with minimal changes applied to make it token-free. ByT5 is available in the same five sizes as mT5 (the details of which are presented in Table 1) and is designed to serve the same purposes as mT5, which is a general-purpose pre-trained multilingual text-to-text model. Experiments conducted in [36] demonstrate the effectiveness of ByT5 models for tasks involving short-to-medium length text sequences, although the authors point out that the time taken for inference and fine-tuning may mean that they are less useful for tasks involving longer sequences.

ByT5 is trained on the same corpus as mT5 [37], the Colossal Cleaned Crawled Corpus (C4). This corpus spans 101 languages and comprises more than 800 GB of data. See Table 2 for more information on the absolute and relative volumes of training data in C4 for the languages studied in this work. The model operates at a byte-level and employs a fixed vocabulary size of 256, representing the available UTF-8 encodings. The vocabulary also has three additional special tokens: padding, end of sentence, and an unused *UNK* token.

Table 1 Details of the available ByT5 models [36]

Size	Params	E/D depth
byt5-small	300M	12/4
byt5-base	582M	18/6
byt5-large	1.2B	36/12
byt5-xl	3.47B	36/12
byt5-xxl	12.9B	36/12

Table 2 Absolute and relative volumes of training data in C4 for the languages studied in this work

Language	Training examples	Romanized training examples	Percentage of corpus
English	3,928,733,379	0	31.00
Japanese[a]	85,226,039	235,000	0.60
Gujarati	1,292,191	0	0.01

[a] For scripted languages where the language has romanized examples, a further split is made to separate the original and the romanized examples

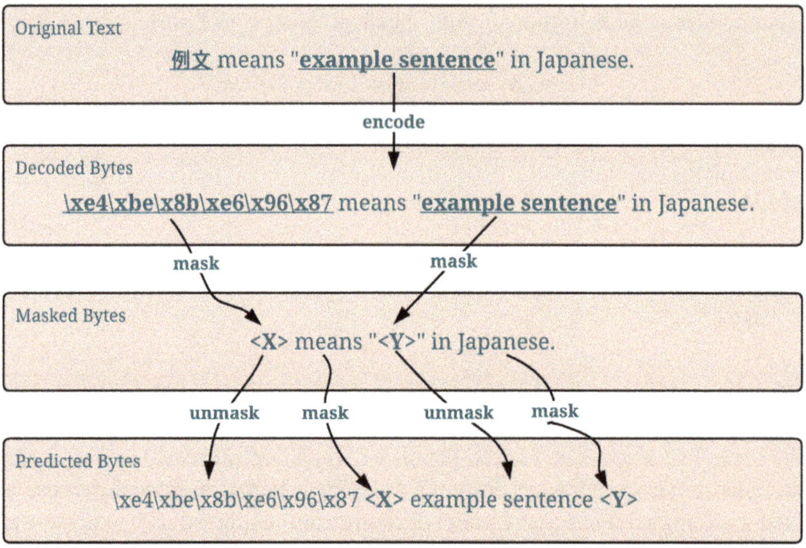

Fig. 2 An illustration of the span corruption training mechanism used in ByT5 [36]

The model utilizes similar training mechanisms to BERT, based on an unsupervised masked language modelling (MLM) technique called span corruption [23]. ByT5 takes the original raw input text and decodes this into bytes, before replacing spans of 20 bytes with a masking token. The goal of the decoder is then to predict the masks for the originally unlabeled text and reconstruct text where the spans are corrupted. See Fig. 2 for an illustration of how this mechanism operates at the byte level.

The authors in [36] found that imbalanced encoder–decoder architectures, where one side of the network has a different depth than the other, can achieve comparable performance to balanced architectures such as those in mT5. This was achieved by decoupling the depth of the encoder–decoder stacks and setting the encoder depth to three times that of the decoder. This modification makes the model similar to encoder-only models like BERT, and however results are improved on both classification and generation tasks in cases where those tasks involve noisy text, such as in TweetQA [35].

4 Experiments

In this section we describe the general conditions under which the experiments for this work were carried out. The details of the new models proposed in this chapter are discussed in Sect. 5 below.

4.1 Languages

The languages studied in this work are English, Japanese, and Gujarati. These languages are selected for comparability with [8] and—as in that work—to provide a diverse sample of languages from different families, including one *scriptio continua* language and one low-resource language. Of further significance to this work in particular is the fact that our English dataset contains only characters represented using a single byte in UTF-8, whereas Japanese and Gujarati consist mainly of three-byte characters, so this language selection enables us to compare the performance of byte-level models across texts with varying per-character byte lengths.

4.2 Datasets

Three of the four datasets from [8] are reused in this work. These are **TedTalks**, which consists of 3997 professional transcriptions of English language Ted Talks, **OshieteQA**, which consists of 42,940 questions and answers crawled from a popular Japanese Q&A site, and **GujaratiNews**, which consists of 3498 news articles crawled from a Gujarati news website. Further details of these datasets are presented in [8]. The same train/test splits are used in this work and the previous work. Only the **TedTalks** dataset was selected for experiments on English in this work as it is larger and more reflective of real-life use cases than the **Brown** dataset used in the previous work, and a direct comparison with [28] was not necessary since the models in that work were already shown to be outperformed by the models in [8].

4.3 Evaluation Metrics

We employ precision, recall, and F-score for each restored feature at the character level as our primary metrics and also report word error rate (WER) for spaced languages only. The details of these metrics are outlined in [8]. As in [8], all metrics were calculated on hypothesis restorations for complete documents from the test set (after recombining chunks generated for model inference) against gold-standard reference versions.

4.4 Models

In this work, we present new fine-tuned ByT5 models that restore spaces, punctuation, and capitalization in a single inference step and are language agnostic. We compare results for these models with those for the only end-to-end, language-agnostic model and the highest-performing pipeline model for English given in [8]. A summary of each of these models is presented below–further details are available in [8].

BiLSTMCharE2E is a character-level BiLSTM sequence-to-sequence classification model. Categorical cross entropy is used for the loss function and softmax is used for the activation function. The following hyperparameters were tuned individually for each dataset by means of a grid search using 10% of the training data with two or three options for each hyperparameter: number of BiLSTM units, batch size, dropout rate, and recurrent dropout rate.

NB + BERTBiLSTM is a pipeline composed of **NB**, a Naive Bayes-based model for space restoration, and **BERTBiLSTM**, a token-level classification model for restoration of capitalization and punctuation with a BERT transformer layer attached to a BiLSTM with sigmoid activation. For **NB**, the following hyperparameters were tuned individually for each dataset by means of a grid search using 5% of the training data with five options for each hyperparameter: L, the maximum possible word length, and λ, the smoothing parameter. For **BERTBiLSTM**, the maximum sequence length was set to 256, and the learning rate, decay, and gradient clip hyperparameters were set to $5e-6$, 0, and 1, respectively.

The details of the fine-tuned ByT5 models proposed for the first time in this work are presented in the next section.

5 Fine-Tuned ByT5 Models

In this section, we present the details of the fine-tuned ByT5 models proposed for restoration of spaces, punctuation, and capitalization.

5.1 Architecture

We fine-tuned two of the five available pre-trained ByT5 models—byt5-small (300M parameters) and byt5-base (582M parameters)—for the restoration of the same features as those restored in [8] for each of the languages included in our study. We use the names **ByT5Small** and **ByT5Base** to refer to our fine-tuned versions of the pre-trained ByT5 models. Model hyperparameters and architecture were kept at the defaults in the simplet5 library for Python,[3] which was otherwise customized for use in this work. The batch size was set to 8, and the AdamW optimizer was used with learning rate 0.0001. For generation, beam search was used with 2 beams, $top\_k = 50$, $top\_p = 0.95$, $repetition\_penalty = 2.5$, and $length\_penalty = 1.0$. Hyperparameter tuning was not carried out for this work but could be explored in future work.

5.2 Preparation of Training Data

Unlike the classification models developed in the previous work, for ByT5 models both source and target sequences are just sequences of byte encodings, which are generated directly from strings of characters. Maximum sequence lengths were set to 256 for both source and target for all languages for our experiments. Our English dataset contains only single-byte characters, so target strings of up to 255 characters were used, whereas Japanese and Gujarati datasets consist primarily of 3-byte characters, so target strings were restricted to 84 characters in order to satisfy the byte length restriction.

Documents in each dataset were concatenated with arbitrary single-byte characters to denote start and end of document (SOD, EOD) since ByT5 models do not specify special tokens for these purposes. We used the Unicode characters U+25B6 ("BLACK RIGHT-POINTING TRIANGLE") and U+25CO ("BLACK LEFT-POINTING TRIANGLE") for SOD and EOD, respectively, after verifying that these characters do not appear in any of our datasets, and removed them from model outputs as part of post-processing at test time. Training target strings were generated by splitting the concatenated documents into strings of the predetermined length for each language, with feature characters removed to generate the corresponding source strings. Average source string lengths (rounded to the nearest integer) were 202 characters/bytes for English, 80 characters/230 bytes for Japanese, and 69 characters/206 bytes for Gujarati.

5.3 Training

Training was carried out on a single NVIDIA GeForce RTX 3090 GPU for the **ByT5Small** models. The larger **ByT5Base** models were found not to fit on a single GPU, so model parallelism was achieved through sharded training across 3 of the same GPUs.[4] Each model was trained for 10 epochs, after which the checkpoint with

[3] https://pypi.org/project/simplet5/.

[4] https://pytorch-lightning.readthedocs.io/en/1.5.10/advanced/advanced_gpu.html.

Table 3 Details of the training checkpoints selected for testing for each model/language combination

Language	Model	Epoch	Training loss	Validation loss
English	**ByT5Small**	5	0.336	0.335
	ByT5Base	5	0.293	0.299
Japanese	**ByT5Small**	6	0.178	0.186
	ByT5Base	8	0.013	0.020
Gujarati	**ByT5Small**	8	0.013	0.016
	ByT5Base	8	0.121	0.145

the lowest validation loss was selected for testing. The details of the checkpoints selected for testing are presented in Table 3. In each case, the lowest validation loss was observed at least 2 epochs before the final epoch, so the models are thought to have converged.

5.4 Post-Processing of Outputs

5.4.1 Chunking

The maximum length of model input sequences was set to 256 bytes, so in order to restore features to test documents containing up to several thousands of characters we used the same chunking method as described in [8]. Chunk lengths were set to 215 for English and 70 for both Japanese and Gujarati. Numbers close to the average length of inputs in the training data for each language were chosen in order to leave space for the outputs to contain all non-feature characters in the input string as well as hypothesized features in the majority of cases (although there is some room for error here, as the character matching function described later in this section ensures that all input characters are reflected in the final output string). A detailed tuning to optimize chunk lengths was not carried out but could be explored in future work.

5.4.2 Unpredictability of Model Outputs

For the classification models developed in [8], output sequences had the same length as input sequences and consisted solely of class labels, from which features could then be inserted in a trivial manner based on class mappings. However, for the generative models studied in this work, there is no guarantee that model outputs contain exactly the characters in the input string, with only hypothesized features inserted. Indeed, while fine-tuning with the available data was sufficient to enforce this behavior for the majority of model outputs, we observed several cases where extra characters that were not present in the input string were inserted in the output string or characters present in the input omitted in the output. Of particular note is

IN: iwasabletotakepartinthenationalyouthconventio</s>
OUT: I was able to take part in the National Youth Convent, I...
I was able to</s>
IN: 100moreafterandthen100moreafterandthen1000moreafterthatbut
OUT: 100 more after, and then 100 more after, and then 1,000 more
after, **and then 1,000 more after** that. But
IN: thinkingoiloiloilwhyisitallhestalkingaboutisoil
OUT: thinking, oil, oil, oil, **oil**. Why is it all hes talking
about is oil.

Fig. 3 Examples of cases of inference for the English **ByT5Small** model where sequences of characters were repeated in the output string more times than they appeared in the input string. Non-feature characters that appear in the input but not in the output are displayed in **bold**. The input string and output strings presented are extracts from actual inference examples from test documents 1, 10, and 11 from the **TedTalks** dataset

the tendency of the models to repeat words or phrases more times than they appear in the input string, which is thought to be a type of oscillatory hallucination [24].

Some examples of this behavior are presented in Fig. 3. In the first example, the model repeats a clause that appeared earlier in the input string beginning with the same letter ("i") as the next character in the input. This occurs close to the end of the input string and is one example of a slight tendency for the model to want to fill out the available space despite having run out of characters in the input string, which may be due to the fact that the end of sequence character `</s>` appears in the final position in the majority of the training examples. In both examples, three repetitions of the same or very similar phrases or words ("and then [NUM] more after" and "oil") prompt the model to output a fourth instance of the same phrase or word, regardless of the next letter in the input (in other words, when they encounter a sequence of the form AAA, the models tend to produce a fourth A). These observations led us to introduce the character matching method described below in Sect. 5.4.3.

5.4.3 Character Matching Method

In our previous work on this topic, we established the expectation that outputs of our restoration models should contain exactly the same non-feature characters as the input text, with only features added (for example, for English, letters in the input can be capitalized and have any combination of a single space/comma/period inserted after them but cannot be inserted or deleted). Apart from being a reasonable requirement for a feature restoration model, our metrics are also based on this expectation, so for comparability with the results in [8] it was necessary to sanitize model outputs so that this requirement was met.

```
def match_chars(input, hypothesis):
    Input: Two strings, input and hypothesis, which are the input string and the
  hypothesis generated through model inference.
    Output: The sanitized hypothesis string that contains exactly the characters in the
  input string, in the same order, with only features inserted.
    output = ''
  /* Start with an empty string */
    for input_char in input:
        match ← RegEx search for first instance of
      pattern /[input_char|input_char.upper()][., ]*/ in hypothesis
      /* e.g. /[aA][., ]*/ */
        if match:
            output ← output + match
            hypothesis ← everything after match in hypothesis
          /* Any extra non-feature characters in hypothesis are skipped
          */
        else:
            output ← output + input_char
          /* Input characters that do not appear in hypothesis are
          inserted without any features */
    return output
```

Fig. 4 Pseudo-code for the character matching function. The RegEx shown is for the case of English and differs depending on the features being restored for each language

We achieved this by defining a character matching function that was applied to the hypothesized output for each chunk immediately after inference and before generating the next chunk. The character matching function takes the input string and the hypothesis generated through model inference and returns a sanitized hypothesis string that contains exactly the characters in the input string, in the same order, with only features inserted. The algorithm for the character matching function is described by the pseudocode in Fig. 4.

The steps described above were found to be sufficient to ensure that all final document outputs satisfied the requirement outlined at the beginning of this subsubsection.

In the implementation of the classification models developed in [8], model outputs (classes) were combined with input strings by applying the features indicated in the output sequence character by character. The character matching function is analogous to that process but takes a different form as the new models developed in this work are generative in nature.

6 Results

In this section, we present the results of inference on the test documents for each dataset using the fine-tuned ByT5 models proposed for the first time in this work for the restoration of spaces, capitalization, and punctuation and compare them with the results presented in [8].

6.1 Overall Performance

The results of the new experiments carried out in this work are presented alongside selected results from [8] in Table 4. We observe that both fine-tuned ByT5 models (**ByT5Small** and **ByT5Base**) outperform the **BiLSTMCharE2E** model proposed in [8] based on the primary metric—overall F-score—for all languages. We conclude from this that knowledge learnt through unsupervised pre-training of byte-level language models can be effectively transferred to the problem of end-to-end restoration of spaces, punctuation, and capitalization through fine-tuning, boosting

Table 4 Comparative results of the experiments carried out in this work and selected results from [8]. In this work, results are presented as percentages rather than floating point numbers to save space—hence a score reported as 0.95 in [8] corresponds to a score of 95 in the below table. The result for the highest-performing model for each metric is displayed in **bold**

English (TedTalks)

	Spaces			CAPS			Periods			Commas			All			
Model	P	R	F	P	R	F	P	R	F	P	R	F	P	R	F	WER
NB + BERTBiLSTM	99	99	99	**88**	81	84	**82**	**82**	**82**	**77**	67	**72**	**96**	**95**	95	**8.49**
BiLSTMCharE2E	98	98	98	69	62	65	58	53	56	50	38	43	92	89	90	20.48
ByT5Small	**100**	**100**	**100**	84	83	83	78	77	77	69	68	68	95	95	95	9.49
ByT5Base	**100**	**100**	**100**	86	**84**	**85**	81	79	80	71	**71**	71	**96**	**95**	**96**	8.52

Japanese (OshieteQA)

	Periods			Commas			Question marks			All		
Model	P	R	F	P	R	F	P	R	F	P	R	F
BiLSTMCharE2E	54	53	54	**31**	31	31	**56**	48	52	**43**	43	43
ByT5Small	**69**	**85**	**76**	26	84	**40**	54	79	**64**	39	84	**53**
ByT5Base	63	84	72	20	**87**	33	43	**82**	57	32	**85**	46

Gujarati (GujaratiNews)

	Spaces			Periods			Commas			All			
Model	P	R	F	P	R	F	P	R	F	P	R	F	WER
BiLSTMCharE2E	97	97	97	86	81	83	57	45	50	96	94	95	13.85
ByT5Small	**98**	**98**	**98**	84	**90**	87	66	55	60	96	**97**	96	9.91
ByT5Base	**98**	**98**	**98**	**87**	89	**88**	**67**	**60**	**64**	**97**	96	**97**	**9.53**

performance compared to a BiLSTM architecture without pre-training that can only learn from the available training data.

Furthermore, in the results for the English language, we observe for the first time a better performance for a token-free model compared to a pipeline containing a token-based model, suggesting that token-free models can deliver state-of-the-art results across all features, with the added advantages that spaces can be restored at the same time as other features in a single inference step. **ByT5Base** outperforms **BERTBiLSTM** for spaces, where pre-trained transformers were not leveraged in **BERTBiLSTM**, and capitalization, where the inability of **BERTBiLSTM** to restore mid-token capitalization is thought to have negatively impacted character-level metrics, while **BERTBiLSTM** still slightly outperforms **ByT5Base** for punctuation restoration, although the differences are very minimal for all features.

A larger jump in performance was observed for English and Japanese (90 → 96 and 52 →64, respectively) compared to Gujarati (95 → 97). This may be due to some extent to the different characteristics of the datasets used for each language, which are commented on in further detail in [8], but could also be due to the much larger volume of training data in pre-training of the ByT5 models for English and Japanese compared to Gujarati (see Table 1 in Sect. 3.3). In general, the improved performance for all languages suggests that a vast amount of pre-training data is not necessary to improve performance. It would be interesting to observe how pre-trained transformer models are compared to those without pre-training for languages for which no data were available for pre-training—however this is left for future work.

6.2 Mid-Token Capitalization and Punctuation

One of the biggest advantages of token-free approaches to restoration of capitalization and punctuation noted in [8] was the ability of token-free models to restore features that do not occur at the beginning or end of a token. The byte-based models proposed in this work perform significantly better in this task compared to the **BiLSTMCharE2E** model proposed in the previous work. For the English dataset, 66% of tokens in the reference test documents that contained mid-token features were correctly reflected in the outputs of the **ByT5Base** model for English, compared to 47% for **BiLSTMCharE2E**. Among the tokens for which mid-token features were correctly restored at least once by **ByT5Base** but never by **BiLSTMCharE2E** are "LinkedIn," "PlayStation," and "www.stopbribes.org."

6.3 Precision vs. Recall

The generative byte-level models proposed for the first time in this work are observed to lean towards higher recall and lower precision compared to the sequence-to-sequence classification models proposed in [8]. For four out of the ten features restored across the three languages, **ByT5Base** produced the highest recall but was outperformed by at least one other model for precision. It is thought that recall for the previously proposed classification models was negatively impacted by class imbalance causing a reduced tendency to produce certain classes in the absence of very strong contextual evidence, whereas this is not an issue for generative models where features are treated as byte sequences with a high relative frequency in the training data. In the case of commas and question marks for Japanese, the imbalance between recall and precision is particularly extreme and negatively impacts on the overall F-score for the generative byte-based models, although they still outperform **BiLSTMCharE2E** overall. The variation between precision and recall across different features for different models suggests that ensemble models combining two or more models could produce very good results for the overall task.

6.4 Effect of Model Size

ByT5Base outperformed **ByT5Small** for English and Gujarati, but not for Japanese. The exception for Japanese is thought to be largely due to the lack of consistency in the usage of punctuation in the Japanese dataset, which was commented upon in [8]. We therefore believe that larger pre-trained models are generally advantageous to the restoration task and hypothesize that further improvement in the results would be observed if fine-tuning were carried out on the remaining three available ByT5 models, although the downside of using these models is that their large size makes training and inference computationally demanding and time-consuming.

6.5 Considerations for Non-English Languages

We noted in Sect. 4.1 our Japanese and Gujarati datasets consist primarily of three-byte characters, whereas the English dataset contains only single-byte characters. This meant that the number of characters in training examples and inference chunks had to be set to three times smaller for non-English languages than for English, reducing the amount of context available for non-English languages. However, we still saw improved results for non-English languages when using pre-trained byte-level models compared to the character-level models proposed in [8], so we conclude that the advantages of pre-training outweigh the reduction in context and that byte-level approaches can be effective for languages in which single characters are encoded as multiple bytes.

In [8], we observed that models trained at the character level were not able to accurately restore features to Gujarati text, the only language in our study for which grapheme clusters are not equivalent to characters. In the previous work, we switched our implementation to process text at the level of grapheme clusters instead of characters in order to enable our models to effectively process Gujarati text, thereby maintaining language agnosticity. This therefore left open the question of whether any special treatment of Gujarati text would be required for the byte-level models proposed in this work. We found that no such special treatment was required and hence conclude that the extra linguistic knowledge provided by the pre-trained transformer models was sufficient to prevent the models from inserting features in the middle of grapheme clusters.

The above two observations lead us to conclude that the byte-level models proposed in this work are language agnostic to at least the same extent as the **BiLSTMCharE2E** model proposed in [8]. Of course, experiments on a wider range of languages—in particular languages for which no pre-training data is included in the pre-trained transformer model—would help to verify this claim, but this is left for future work.

7 Conclusion and Future Work

We addressed the task of restoration of spaces, capitalization, and punctuation to unformatted sequences of characters, which requires processing of text at a level below the token since token boundaries cannot be ascertained directly from input texts. This work takes the character-level sequence-to-sequence classification model proposed in [8] one step further by proposing a new generative approach to the restoration task employing pre-trained byte-level language models. A character matching function was required to ensure that all characters in the input texts were reliably mapped to output texts, and training had to be distributed across multiple GPUs in order to leverage the larger base versions of the ByT5 models employed. However, the results were marked improvements in overall F-scores for all three languages studied. Furthermore, the best byte-level model for English outperformed the highest-performing pipeline model from [8], indicating that token-level processing is not required in order to achieve state-of-the-art results for the task considered.

The observation that the previously proposed models continue to outperform the newly proposed byte-level models for some metrics on a subset of features, in particular tending to achieve higher precision, leads us to hypothesize that combining the outputs of two or more of the models studied using ensemble techniques could lead to better overall performance. This is left for future work. The results of this study—which employed a selection of languages with diverse linguistic qualities and varying levels of pre-training data—suggest that the byte-level approach is viable for a wide range of languages, and however further work studying a greater number of languages would help to cast light on the relationship between training data volume and model performance and the language agnosticity of the proposed approach in general.

References

1. Al-Rfou, R., Choe, D., Constant, N., Guo, M., Jones, L.: Character-level language modeling with deeper self-attention. Proc. AAAI Confer. Artif. Intell. **33**(01), 3159–3166 (2019). https://doi.org/10.1609/aaai.v33i01.33013159. https://ojs.aaai.org/index.php/AAAI/article/view/4182
2. Bakare, A.M., Anbananthen, K.S.M., Muthaiyah, S., Krishnan, J., Kannan, S.: Punctuation restoration with transformer model on social media data. Appl. Sci. **13**(3), 1685 (2023). https://doi.org/10.3390/app13031685. https://www.mdpi.com/2076-3417/13/3/1685
3. Chay-intr, T., Kamigaito, H., Okumura, M.: Character-based Thai word segmentation with multiple attentions. In: Proceedings of the International Conference on Recent Advances in Natural Language Processing (RANLP 2021), pp. 264–273. INCOMA Ltd., Held Online (2021). https://aclanthology.org/2021.ranlp-1.31
4. Conneau, A., Khandelwal, K., Goyal, N., Chaudhary, V., Wenzek, G., Guzmán, F., Grave, E., Ott, M., Zettlemoyer, L., Stoyanov, V.: Unsupervised cross-lingual representation learning at scale. Computing Research Repository **arXiv:911.02116** (2019). http://arxiv.org/abs/1911.02116. Version 2
5. Courtland, M., Faulkner, A., McElvain, G.: Efficient automatic punctuation restoration using bidirectional transformers with robust inference. In: Proceedings of the 17th International Conference on Spoken Language Translation, pp. 272–279. Association for Computational Linguistics, Online (2020). https://doi.org/10.18653/v1/2020.iwslt-1.33. https://aclanthology.org/2020.iwslt-1.33
6. Devlin, J., Chang, M.W., Lee, K., Toutanova, K.: BERT: Pre-training of deep bidirectional transformers for language understanding. In: Proceedings of the 2019 Conference of the North American Chapter of the Association for Computational Linguistics: Human Language Technologies, Volume 1 (Long and Short Papers), pp. 4171–4186. Association for Computational Linguistics, Minneapolis, Minnesota (2019). https://doi.org/10.18653/v1/N19-1423. https://aclanthology.org/N19-1423
7. Duan, S., Zhao, H.: Attention is all you need for Chinese word segmentation. In: Proceedings of the 2020 Conference on Empirical Methods in Natural Language Processing (EMNLP), pp. 3862–3872. Association for Computational Linguistics, Online (2020). https://doi.org/10.18653/v1/2020.emnlp-main.317. https://aclanthology.org/2020.emnlp-main.317
8. Dyer, L., Hughes, A., Shah, D., Can, B.: Comparison of token- and character-level approaches to restoration of spaces, punctuation, and capitalization in various languages. In: Proceedings of the 5th International Conference on Natural Language and Speech Processing (ICNLSP 2022), pp. 168–178. Association for Computational Linguistics, Trento, Italy (2022). https://aclanthology.org/2022.icnlsp-1.19
9. Guan, Y.: End to end ASR system with automatic punctuation insertion. Computing Research Repository **arXiv:2012.02012** (2020). https://arxiv.org/abs/2012.02012
10. Guerreiro, N.M., Rei, R., Batista, F.: Towards better subtitles: A multilingual approach for punctuation restoration of speech transcripts. Expert Syst. Appl. **186**, 115740 (2021). https://doi.org/10.1016/j.eswa.2021.115740. https://www.sciencedirect.com/science/article/pii/S0957417421011180
11. Guhr, O., Schumann, A.K., Bahrmann, F., Böhme, H.J.: FullStop: Multilingual deep models for punctuation prediction (2021). http://ceur-ws.org/Vol-2957/sepp_paper4.pdf
12. Gupta, A., Chhimwal, N., Dhuriya, A., Gaur, R., Shah, P., Chadha, H.S., Raghavan, V.: indic-punct: An automatic punctuation restoration and inverse text normalization framework for Indic languages. **arXiv:2203.16825** (2022). https://doi.org/10.48550/ARXIV.2203.16825. https://arxiv.org/abs/2203.16825. [Submitted to InterSpeech 2022]
13. Higashiyama, S., Utiyama, M., Sumita, E., Ideuchi, M., Oida, Y., Sakamoto, Y., Okada, I.: Incorporating word attention into character-based word segmentation. In: Proceedings of the 2019 Conference of the North American Chapter of the Association for Computational

Linguistics: Human Language Technologies, Volume 1 (Long and Short Papers), pp. 2699–2709. Association for Computational Linguistics, Minneapolis, Minnesota (2019). https://doi.org/10.18653/v1/N19-1276. https://aclanthology.org/N19-1276
14. Jones, D., Wolf, F., Gibson, E., Williams, E., Fedorenko, E., Reynolds, D., Zissman, M.: Measuring the readability of automatic speech-to-text transcripts. In: 8th European Conference on Speech Communication and Technology (Eurospeech 2003), pp. 1585–1588 (2003). https://doi.org/10.21437/Eurospeech.2003-463. URL https://www.isca-speech.org/archive/pdfs/eurospeech_2003/jones03_eurospeech.pdf
15. Kakwani, D., Kunchukuttan, A., Golla, S., N.C., G., Bhattacharyya, A., Khapra, M.M., Kumar, P.: IndicNLPSuite: Monolingual corpora, evaluation benchmarks and pre-trained multilingual language models for Indian languages. In: Findings of the Association for Computational Linguistics: EMNLP 2020, pp. 4948–4961. Association for Computational Linguistics, Online (2020). https://doi.org/10.18653/v1/2020.findings-emnlp.445. https://aclanthology.org/2020.findings-emnlp.445
16. Liu, J., Wu, F., Wu, C., Huang, Y., Xie, X.: Neural Chinese word segmentation with dictionary. Neurocomputing **338**, 46–54 (2019). https://doi.org/10.1016/j.neucom.2019.01.085. https://www.sciencedirect.com/science/article/pii/S0925231219301286
17. Liu, Y., Ott, M., Goyal, N., Du, J., Joshi, M., Chen, D., Levy, O., Lewis, M., Zettlemoyer, L., Stoyanov, V.: RoBERTa: A robustly optimized BERT pretraining approach. Computing Research Repository **arXiv:1907.11692** (2019). https://arxiv.org/abs/1907.11692
18. Maimaiti, M., Liu, Y., Zheng, Y., Chen, G., Huang, K., Zhang, J., Luan, H., Sun, M.: Segment, mask, and predict: Augmenting Chinese word segmentation with self-supervision. In: Proceedings of the 2021 Conference on Empirical Methods in Natural Language Processing, pp. 2068–2077. Association for Computational Linguistics, Online and Punta Cana, Dominican Republic (2021). https://doi.org/10.18653/v1/2021.emnlp-main.158. https://aclanthology.org/2021.emnlp-main.158
19. Mayhew, S., Tsygankova, T., Roth, D.: NER and POS when nothing is capitalized. Computing Research Repository **arXiv:2012.02012** (2019). https://arxiv.org/abs/1903.11222. Version 2
20. Nguyen, B., Nguyen, V.B.H., Nguyen, H., Phuong, P.N., Nguyen, T.L., Do, Q.T., Mai, L.C.: Fast and accurate capitalization and punctuation for automatic speech recognition using transformer and chunk merging. In: 2019 22nd Conference of the Oriental COCOSDA International Committee for the Co-ordination and Standardisation of Speech Databases and Assessment Techniques (O-COCOSDA), pp. 1–5. IEEE, Cebu, Philippines (2019). https://doi.org/10.1109/O-COCOSDA46868.2019.9041202. https://arxiv.org/abs/1908.02404
21. Peitz, S., Freitag, M., Mauser, A., Ney, H.: Modeling punctuation prediction as machine translation. In: International Workshop on Spoken Language Translation, pp. 238–245. San Francisco, California (2011). https://aclanthology.org/2011.iwslt-papers.7
22. Qiu, Q., Xie, Z., Ma, K., Tian, M.: BERTCWS: unsupervised multi-granular Chinese word segmentation based on a BERT method for the geoscience domain. Ann. GIS **0**(0), 1–13 (2023). https://doi.org/10.1080/19475683.2023.2186487
23. Raffel, C., Shazeer, N., Roberts, A., Lee, K., Narang, S., Matena, M., Zhou, Y., Li, W., Liu, P.J.: Exploring the limits of transfer learning with a unified text-to-text transformer. J. Mach. Learn. Res. **21**(1), 5485–5551 (2020). https://jmlr.org/papers/volume21/20-074/20-074.pdf
24. Raunak, V., Menezes, A., Junczys-Dowmunt, M.: The curious case of hallucinations in neural machine translation. In: Proceedings of the 2021 Conference of the North American Chapter of the Association for Computational Linguistics: Human Language Technologies, pp. 1172–1183. Association for Computational Linguistics, Online (2021). https://doi.org/10.18653/v1/2021.naacl-main.92. https://aclanthology.org/2021.naacl-main.92
25. Rizqullah, R.D., Budi, I.: Text normalization on code-mixed Twitter text using language detection. In: 2022 Seventh International Conference on Informatics and Computing (ICIC), pp. 1–4 (2022). https://doi.org/10.1109/ICIC56845.2022.10006949. https://scholar.ui.ac.id/en/publications/text-normalization-on-code-mixed-twitter-text-using-language-dete

26. Samuel, D., Straka, M.: ÚFAL at MultiLexNorm 2021: Improving multilingual lexical normalization by fine-tuning ByT5. In: Proceedings of the Seventh Workshop on Noisy User-generated Text (W-NUT 2021), pp. 483–492. Association for Computational Linguistics, Online (2021). https://doi.org/10.18653/v1/2021.wnut-1.54. https://aclanthology.org/2021.wnut-1.54
27. Shao, Y., Hardmeier, C., Nivre, J.: Universal word segmentation: Implementation and interpretation. Trans. Assoc. Comput. Linguist. **6**, 421–435 (2018). https://doi.org/10.1162/tacl_a_00033. https://aclanthology.org/Q18-1030
28. Sivakumar, J., Muga, J., Spadavecchia, F., White, D., Can, B.: A GRU-based pipeline approach for word-sentence segmentation and punctuation restoration in English. In: 2021 International Conference on Asian Language Processing (IALP), pp. 268–273 (2021). https://doi.org/10.1109/IALP54817.2021.9675269. https://ieeexplore.ieee.org/abstract/document/9675269
29. Stankevičius, L., Lukočevičius, M., Kapočiūtė, J., Briedienė, M., Krilavičius, T.: Correcting diacritics and typos with a ByT5 transformer model. Appl. Sci. **12**(5) (2022). https://doi.org/10.3390/app12052636. https://www.mdpi.com/2076-3417/12/5/2636
30. Susanto, R.H., Chieu, H.L., Lu, W.: Learning to capitalize with character-level recurrent neural networks: An empirical study. In: Proceedings of the 2016 Conference on Empirical Methods in Natural Language Processing, pp. 2090–2095. Association for Computational Linguistics, Austin, Texas (2016). https://doi.org/10.18653/v1/D16-1225. https://aclanthology.org/D16-1225
31. Sutskever, I., Vinyals, O., Le, Q.V.: Sequence to sequence learning with neural networks (2014). https://arxiv.org/abs/1409.3215
32. Tan, S., Behre, P., Kibre, N., Alphonso, I., Chang, S.: Four-in-One: a joint approach to inverse text normalization, punctuation, capitalization, and disfluency for automatic speech recognition. In: 2022 IEEE Spoken Language Technology Workshop (SLT), pp. 677–684 (2023). https://doi.org/10.1109/SLT54892.2023.10023257. https://arxiv.org/abs/2210.15063
33. Vaswani, A., Shazeer, N., Parmar, N., Uszkoreit, J., Jones, L., Gomez, A.N., Kaiser, Ł., Polosukhin, I.: Attention is all you need. Adv. Neural Inf. Process. Syst. **30**, (2017). https://arxiv.org/abs/1706.03762
34. Wang, F., Chen, W., Yang, Z., Xu, B.: Self-attention based network for punctuation restoration. In: 24th International Conference on Pattern Recognition (ICPR), pp. 2803–2808. IEEE, Beijing, China (2018). https://doi.org/10.1109/ICPR.2018.8545470. https://ieeexplore.ieee.org/document/8545470
35. Xiong, W., Wu, J., Wang, H., Kulkarni, V., Yu, M., Chang, S., Guo, X., Wang, W.Y.: TWEETQA: A social media focused question answering dataset. In: Proceedings of the 57th Annual Meeting of the Association for Computational Linguistics, pp. 5020–5031. Association for Computational Linguistics, Florence, Italy (2019). https://doi.org/10.18653/v1/P19-1496. https://aclanthology.org/P19-1496
36. Xue, L., Barua, A., Constant, N., Al-Rfou, R., Narang, S., Kale, M., Roberts, A., Raffel, C.: ByT5: Towards a token-free future with pre-trained byte-to-byte models. Trans. Assoc. Comput. Linguist. **10**, 291–306 (2022). https://doi.org/10.1162/tacl_a_00461. https://aclanthology.org/2022.tacl-1.17
37. Xue, L., Constant, N., Roberts, A., Kale, M., Al-Rfou, R., Siddhant, A., Barua, A., Raffel, C.: mT5: A massively multilingual pre-trained text-to-text transformer. In: Proceedings of the 2021 Conference of the North American Chapter of the Association for Computational Linguistics: Human Language Technologies, pp. 483–498. Association for Computational Linguistics, Online (2021). https://doi.org/10.18653/v1/2021.naacl-main.41. https://aclanthology.org/2021.naacl-main.41
38. Yi, J., Tao, J., Bai, Y., Tian, Z., Fan, C.: Adversarial transfer learning for punctuation restoration. Computing Research Repository **arXiv:2004.00248** (2020). https://arxiv.org/abs/2004.00248
39. Zhang, Q., Liu, X., Fu, J.: Neural networks incorporating dictionaries for Chinese word segmentation. Proc. AAAI Confer. Artif. Intell. **32**(1) (2018). https://doi.org/10.1609/aaai.v32i1.11959. https://ojs.aaai.org/index.php/AAAI/article/view/11959

Hierarchical Multi-task Learning with Articulatory Attributes for Cross-Lingual Phoneme Recognition

Kevin Glocker and Munir Georges

1 Introduction

Despite the introduction of several highly effective architectures for speech recognition in recent years, most of them require a significant amount of training data for each language. Unfortunately, for many languages spoken around the world, corpora with annotated speech recordings either contain few utterances or are unavailable.

To address this issue, (pre-)training end-to-end architectures on large multilingual corpora from high-resource languages has been explored. In [25], the authors fine-tune a multilingually pretrained wav2vec 2.0 model for phoneme recognition to tackle the cross-lingual transfer task. Such ASR models can either be fine-tuned for low-resource languages with limited training data as proposed by, e.g., [21] or directly applied zero-shot without any training data in the target language, as evaluated in [13].

Articulatory attribute systems developed by linguists have been leveraged in several ASR architectures to improve phoneme recognition performance. In [12], signature matrices are used to map distribution over articulatory attributes to phoneme distributions. In other approaches, attributes are used as inputs in the form of trainable embeddings. In such systems, attribute embeddings can be either assigned to each attribute individually as proposed by, e.g., [13] or computed as linear or nonlinear transformations of whole attribute vectors as in, e.g., [27].

K. Glocker (✉)
AImotion Bavaria, Technische Hochschule Ingolstadt, Ingolstadt, Germany
e-mail: kevin.glocker@thi.de

M. Georges
AImotion Bavaria, Technische Hochschule Ingolstadt, Ingolstadt, Germany
Intel Labs, Munich, Germany
e-mail: munir.georges@thi.de

M. Abbas (ed.), *Practical Solutions for Diverse Real-World NLP Applications*,
Signals and Communication Technology,
https://doi.org/10.1007/978-3-031-44260-5_4

Articulatory attributes have also been included in ASR systems in the form of multi-task learning as proposed by [11]. In their architecture for Mandarin ASR, separate articulatory attribute classifiers and triphone states are trained at the same time using shared layers with a time-delay neural network architecture on forced alignments.

This work is an extension of our previous paper on hierarchical multi-task learning for cross-lingual phoneme recognition [5]. Hierarchical multi-task learning is used to jointly learn the recognition of articulatory attributes and phonemes. The key difference to regular multi-task learning is the addition of a direct connection between the attribute classifiers and the phoneme classifier. In this work, we provide a corrected and more detailed evaluation of the architectures introduced in our previous work and present evaluation results on a language and phoneme level. Furthermore, we propose an additional modification to our hierarchical multi-task architecture. The evaluation confirms that this method substantially improves performance by solving an increase in phoneme deletion errors over the baseline.

The proposed architectures for phoneme recognition are introduced in Sect. 2. The details about the models, datasets, and training process used for evaluation are given in Sect. 3. Evaluation results are then analyzed and models compared in Sect. 4 in the supervised and zero-shot cross-lingual settings. Finally, we conclude in Sect. 5.

2 Phoneme Recognition Architecture

Our approach uses a hybrid convolution and transformer acoustic model architecture which is described in Sect. 2.1. In Sect. 2.2 we then introduce our architectures for hierarchical multi-task classification of phonemes and articulatory attributes.

2.1 Hybrid Transformer Acoustic Model

The proposed architecture for cross-lingual phoneme recognition is illustrated in Fig. 1. To encode sequences of acoustic frames, a hybrid convolution and transformer architecture is used. We derive the architecture of this model and the choices of hyperparameters from a transformer acoustic model proposed by [22]. Out of their introduced transformer architectures, we base our acoustic model on the variant trained with connectionist temporal classification (CTC) [6].

To encode utterances with our model, we first resample input audio to 16 kHz. As our acoustic features, we extract 25 ms frames with a stride of 10 ms and compute 40-dimensional MFCC features. As the initial layers of our models, we use two Gated Linear Unit (GLU) [4] convolutions. The MFCCs from our preprocessing step are then passed through those layers to encode local context. The kernel size of both GLU layers is three frames. The layers use 512 channels and 400 channels, respectively. We regularize the convolution layers by applying layer normalization

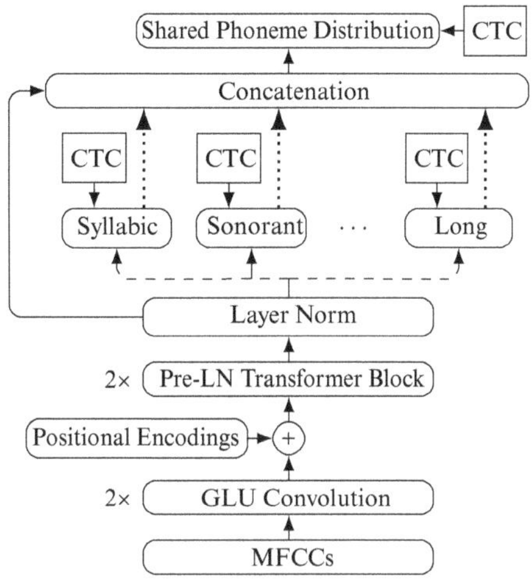

Fig. 1 Illustration of the hybrid convolutional transformer phoneme recognition architecture, including the attribute and phoneme classifiers. The dashed connections between the layer norm and attribute classifiers are used in all multi-task architectures. The dotted connections between attribute classifiers and the phoneme classifier are only used in hierarchical multi-task learning

to their inputs and dropout to their outputs. To increase the receptive field of the convolution part of the architecture to 5 frames, the second GLU layer uses a stride of two.

For global context, we use a shallow two-layer transformer. To enable the transformer to learn positional information from the encoded acoustic sequence, we add sinusoidal positional encodings [23] to the output hidden vectors of the convolution layers. Our main modification compared to [22] is that we use Pre-LN transformer blocks proposed by [24]. Following their findings, we train without warmup. Motivated by [23], the feedforward layers of each block have a hidden size of 2048 and four heads are used in the multi-head attention layers. We set the dropout rate to 0.2 throughout the network.

2.2 *Hierarchical Multi-task Classification of Phonemes and Attributes*

In the previous work on multi-task learning with articulatory attributes, classifiers were trained independently [11]. In contrast, we enable a direct flow of information between attributes and phonemes by training the classifiers in a hierarchical structure. Previously, using such cascading information has also been successfully applied to optimizing tasks that depend on each other jointly, such as part-of-speech tagging and dependency parsing [3].

We follow regular multi-task learning by optimizing each classification layer independently but simultaneously using CTC. The normalized output of the transformer acoustic model is used as the input to both the attribute classifiers and the phoneme classifier. To pass information from the attribute classifiers directly to the phoneme classifier, the probability distributions from all articulatory attribute classifiers are additionally concatenated together with the transformer output at each frame. The resulting vectors are then used as the inputs to the phoneme classifier. More formally, for each time step t given a set of attribute classifier logits A_t, the transformer hidden vector h_t, and the weights and biases of the phoneme projection layer W and b, the phoneme logits p_t are computed as follows:

$$v_t = \left(\bigoplus_{a \in A_t} \text{softmax}(a)\right) \oplus h_t \tag{1}$$

$$p_t = W^T v_t + b \tag{2}$$

A direct one-to-one correspondence is kept between attribute and phoneme labels during training. To achieve this, articulatory attribute vectors are directly mapped to each phoneme without merging repetitions. In this work, the hierarchy between the attribute and phoneme classifiers is flat, with only the articulatory attribute and phoneme level. However, the introduced method of passing probability distributions from one classifier to another generalizes to any directed acyclic graph representing articulatory attribute structures.

Probabilities of attributes tend to form single frame spikes in CTC, with most remaining frames being dominated by high blank probabilities [6]. Since each attribute classifier is independently supervised with CTC, it is possible that different attribute classifiers produce spikes at different frames. Given such an imperfect alignment, it is possible that in a given frame, the phoneme classifier only receives spikes of high attribute probabilities from some attribute classifiers and only very high blank probabilities from others.

Motivated by this assumption, a novel method is proposed in this chapter. Blank logits are removed before computing the attribute probability distributions that are passed as inputs to the phoneme classifier. More formally, consider the attribute logits $a = [a_1, \ldots, a_N]$, where $a \in A_t$ and N is the number of possible values for each attribute including the blank label. Given that blank logits are always the first component a_1 in each logit vector a, the blank logits are removed from each vector through slicing:

$$A'_t = \{a \in A_t \mid [a_2, \ldots, a_N]\} \tag{3}$$

A new architecture variant is defined which is named "No Blanks" in this chapter. In this architecture, the softmax in Equation 1 is computed using the logit vectors from A'_t instead of A_t. With this change, attribute probabilities no longer change dramatically in magnitude between frames with spikes and those where blank probabilities would otherwise dominate. It also slightly reduces the total number of required parameters in the linear layer used for phoneme classification by $2|A'|$. This includes one parameter for the weight and one for the bias of each classification layer.

3 Evaluation

The proposed hierarchical multi-task architectures for cross-lingual phoneme recognition with and without blank removal, regular multi-task learning, and a baseline are evaluated in this section. In total, four different models are evaluated:

1. "Phonemes Only" serves as a baseline. In this architecture, only the phoneme classifier is used. Attribute information is only applied to phoneme mapping at test time.
2. "Multi-Task" uses regular multi-task learning, where the phoneme classifier does not receive attribute probabilities as an input.
3. "Hierarchy" uses hierarchical multi-task learning as introduced in Sect. 2.2.
4. "Hierarchy No Blanks" uses the "No Blanks" variant of the hierarchical multi-task architecture.

The datasets for training and evaluation are described in Sect. 3.1. Section 3.2 contains details of model training, including hyper-parameters, software libraries, and hardware.

3.1 Datasets

The crowdsourced recordings of sentences from the Mozilla Common Voice corpus [1] (Version 10.0) were used for both training and evaluation in the supervised setting. The sentences are split into tokens using the Stanza tool [20], followed by punctuation removal. Each token is then independently transcribed into phonemes using the Epitran tool [17]. All rare characters that Epitran cannot transcribe into the international phonetic alphabet are removed beforehand. Finally, the transcriptions are divided into the phoneme segments that are available in the Panphon database [18]. From the training data of each language, we extract phoneme inventories by taking the union over all phonemes occurring in their transcriptions. The inventories were then used for both training and evaluation. For the phoneme classifier, a shared phoneme inventory is then constructed by taking the union of all phonemes from the inventories of each language similar to [25]. We include one classifier for each of the 24 articulatory attributes from Panphon in our multi-task models and use the attributes from the database for each phoneme for supervising the attribute classifiers.

At most 15,000 sentences from the training sets of 24 languages from Common Voice were used for training the multilingual phoneme recognizer. These languages were chosen since both a Stanza tokenizer and an Epitran model are available for them. No changes were made to the original development and test sets.

For evaluating our models on zero-shot transfer, the first release[1] of the UCLA Phonetic Corpus (UCLA) published by [15] was used as in [13]. The corpus covers 95 low-resource languages from five continents with a total of 5,509 validated utterances with phoneme transcriptions. To ensure that the evaluation on this corpus only tests the cross-lingual transfer capabilities of our models, recordings for Czech, Dutch, Maltese, Hindi, and Hungarian were removed since these languages are also included in the training data. Tones were not included in training or considered during evaluation since the focus of this work is on phoneme recognition. Results on UCLA are not affected by this choice since its transcriptions do not contain tones.

The "tr2tgt" approach introduced by [25] is used to handle different inventories and OOV phonemes in the test languages. In this approach, the minimum Hamming distance between attribute vectors is used to map phonemes from the shared inventory predicted by the model to each target inventory. The inventory files included in UCLA are used for this mapping in the cross-lingual setting even if they include a phoneme that does not appear in a transcription.

The inventories in UCLA also contain some complex phoneme segments such as diphthongs and multiple articulations that are missing from the Panphon database. In these cases, attribute vectors were obtained approximately by taking the attribute vector of the longest possible phoneme prefix of the complex segment that exists in the database and assigning it to the full segment. For instance, the Maasina Fulfulde language uses the complex segment [ŋg], for which no single entry exists in Panphon. In this case, the longest prefix we can use as an approximation is [ŋ]. The resulting approximate attribute vectors are used both for mapping phonemes at inference time and for evaluating the attribute classifiers. In future work, the use of another phoneme database that already contains more accurate attributes for complex segments such as PHOIBLE [16] can be evaluated for this purpose.

3.2 Training

The batch sizes are selected dynamically for computational efficiency. For each batch, acoustic feature sequences are randomly sampled uniformly until the product of the batch and frame sequence dimensions reaches 320,000. The Adam optimizer [9] was used for training with $\beta_1 = 0.9$ and $\beta_2 = 0.98$ as in [23]. A learning rate of 0.001 was used. The training was stopped once the average validation set losses did not decrease for more than 3 epochs.

The transformer acoustic model was implemented in the PyTorch framework [19], using Torchaudio [26] for audio processing and feature extraction. All models were trained on single MIG instances with 20 GB of memory of an NVIDIA A100 80 GB GPU. Models using multi-task learning took 37 hours and 35 minutes on average to train until convergence as determined by the early stopping criterion.

[1] https://github.com/xinjli/ucla-phonetic-corpus/releases/tag/v1.0

The baseline without the attribute-level classifiers took 21 hours and 31 minutes to train.

4 Discussion

The phoneme error rate (PER) and attribute error rate (AER) for each model are used as the evaluation metrics. The AER is computed for each attribute individually and then averaged over all attributes into a single score. Furthermore, the insertion, deletion, and substitution rates are reported to gain further insights into the performance characteristics of our models. Note that due to a previously undetected but now resolved bug in the selection of phoneme mappings and inventories across languages, our results differ from those originally reported in [5]. This issue affected the results on both test sets. Results on UCLA were affected the most since there is more of a difference between the shared phoneme inventory of each model taken from the training languages and the inventories of the individual languages than on Common Voice.

The models are compared on languages seen during training and unseen low-resource languages in Sects. 4.1 and 4.2, respectively, followed by combined analysis across both datasets in Sect. 4.3. Finally, the phoneme-level errors are investigated in Sect. 4.4.

4.1 Common Voice

The overall performance on phoneme and articulatory attribute recognition on Common Voice can be seen in Table 1. Compared to the "Phonemes Only" model, "Multi-Task" performs almost identically with an increase of less than 0.1 pp. PER. This suggests that sharing hidden representations between both tasks did not improve them in a way that benefits phoneme recognition. The "Hierarchy" model also performs slightly worse than both the baseline by 0.46 pp. PER and the "Multi-Task" model by 0.37 pp. Additionally, the connection between attribute

Table 1 Average phoneme and attribute error rates, their variances between languages, and phoneme insertion, deletion, and substitution rates for the Common Voice test set

Name	PER	PER σ^2	Insertions	Deletions	Substitutions	AER	AER σ^2
Phoneme Only	52.16%	87.42	1.23%	24.74%	26.19%	–	–
Multi-task	52.25%	83.59	**0.81%**	28.65%	**22.79%**	26.49%	31.56
Hierarchy	52.62%	87.52	1.12%	26.31%	25.20%	28.48%	44.23
Hierarchy No Blanks	**49.00%**	127.44	2.80%	**16.36%**	29.84%	**20.52%**	30.74

and phoneme classifiers also leads to an increase in AER by 1.99 pp. In contrast, “Hierarchy No Blanks” outperforms the baseline by 3.16 pp. PER. This shows that providing attribute-level information to the phoneme classifier through the hierarchical connection can improve recognition performance when blank logits are omitted beforehand. Furthermore, the AER is also greatly reduced by 7.96 and 5.97 pp. PER compared to “Hierarchy No Blanks” and “Multi-Task,” respectively.

Further insights can be gained by analyzing the rates of insertion, deletion, and substitution errors separately. Overall, most of the errors across models come from substitutions and deletions, with insertions not exceeding 2.8%. The low number of insertion errors is likely due to our use of downsampling in the initial GLU layers of the acoustic model. The length of predicted phoneme sequences is constrained to at most the length of the input sequence. This means that shorter input sequences leave fewer possible frames for insertion errors to occur in. False positives are therefore more likely to be substitution errors.

While the overall PERs are nearly identical between “Phoneme Only,” “Multi-Task,” and “Hierarchy,” particularly deletions and substitution errors vary much more across models. When comparing the baseline to “Multi-Task” and “Hierarchy,” the latter models produce a lower number of substitution errors but yield more deletion errors instead. This increase in deletions cancels out the reduction in substitutions, leading to the PERs between models remaining similar. In contrast, the deletion rate in “Hierarchy No Blanks” is substantially lower with a reduction of 8.38 pp. compared to the baseline. The increase of substitution errors by 3.65 pp. indicates that many of the phonemes that are no longer deleted are still incorrectly predicted. However, the increase in the number of decoded phonemes overall leads to more correct predictions and an overall lower PER. The overall pattern of very few insertions and a large number of deletions has also been observed in the models proposed for cross-lingual phoneme recognition in [14].

A potential explanation for the reduction of deletions after removing blank labels is that different attribute classifiers producing high blank probabilities and spikes at different frames no longer have an effect on the probability distributions received by the phoneme classifier. It is conceivable that receiving high blank probabilities from multiple not perfectly aligned attribute classifiers could lead to more phoneme deletions. However, the reduction of deletions after the addition of the hierarchical connection from “Multi-Task” to “Hierarchy” suggests that this effect does not occur in practice. It is therefore more likely that the improvement of both attribute and phoneme recognition primarily stems from the more consistent magnitudes of the attribute probabilities across frames after blanks are removed.

When considering the variances of PER (PER σ^2) and AER (AER σ^2), AERs are substantially more stable across languages. This may be explained by attributes being less sparse and occurring in more languages than individual phonemes. Future work could analyze this in more detail. The AER variances are very similar in “Multi-Task” and “Hierarchy,” while the variance of “Hierarchy” is around 40% higher than in the other two models. This further emphasizes the issues of negative transfers to attribute classification when using a hierarchical connection without blank removal, which completely negates this increase in AER variance. Although

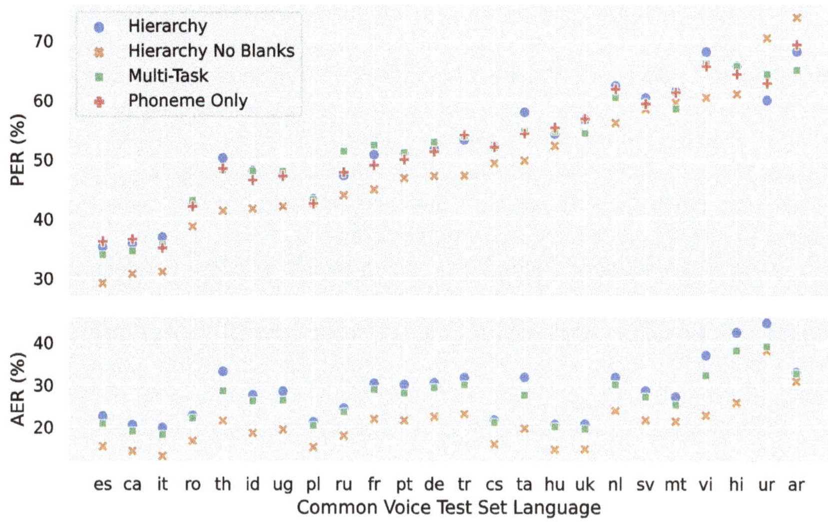

Fig. 2 PER and AER on the test sets from Common Voice for each of the languages used for training. The PERs are shown for all models. No AERs are plotted for "Phoneme Only," as the baseline architecture does not contain attribute classifiers

the overall PERs are substantially lower in "Multi-Task No Blanks" than all other models, the variance of PERs between languages is 46% higher than for the baseline, warranting further investigation on the language level.

Figure 2 shows the PERs and AERs for the test sets of each of the languages from Common Voice that were used for training. For all models, Arabic and Urdu are among the languages with the highest PER. In contrast to the results of "Hierarchy No Blanks" on most other languages, the model yields PERs for the two languages that are substantially higher than those from other models. This also makes Arabic and Urdu the languages that are by far the worst recognized by "Hierarchy No Blanks." A major limitation in evaluating both languages is that transcriptions are written in Arabic script without vowel marks. For this reason, Epitran cannot transcribe short vowels in these languages [17].

However, we find that our models still learn to recognize some of the short vowels in both languages from the transcriptions of other languages. This is also reflected in the insertions rates in both languages, which are much higher than average at 18.9% and 18% for Arabic and Urdu, respectively, and consist mostly of vowel insertion errors. The two languages are also the only languages in which the "No Blank" model performs worse than "Hierarchy." With the assumption that more insertions of vowels lead to more accurate transcriptions than the reference, the increase in PER through more insertions could be a sign of better generalization across languages. Conclusively proving this requires further analysis using more accurate transcriptions from different corpora or grapheme-to-phoneme systems in future work.

Similar effects of orthographic depth on phoneme recognition performance have also been found in [14]. In particular, they attribute the high PER for Swedish in their model to its deep orthography. In our results, Swedish is the language with the sixth highest PER in our data, likely for the same reason. Vietnamese has the fourth highest PER in the test set. This is probably because it is the second lowest resource language in the training data with only 2259 validated utterances. It is also the only Austroasiatic language in the training data and can therefore not benefit from the presence of closely related languages in the corpus.

In contrast, phoneme recognition is the most accurate for the four romance languages Spanish, Catalan, Italian, and Romanian. They likely benefit the most from the multilingual settings since they are closely related. Another probable factor is the orthographic depth of the languages. In particular, Italian has a more shallow orthography than French [2]. This could explain the substantially French PER (13.8 pp.) in "Hierarchical No Blanks" despite also being a romance language.

4.2 UCLA Phonetic Corpus

The results of applying our models to cross-lingual phoneme and articulatory attribute recognition on UCLA are presented in Table 2. Compared to supervised results, cross-lingual transfer capabilities differ much more between models. We find that the "Multi-Task" model in this setting yields a PER that is 4.45 pp. higher than the baseline. This suggests that regular multi-task learning suffers from negative transfer effects between attribute classifiers and the phoneme classification layer. In contrast to the results on Common Voice, the "Hierarchy" model outperforms "Multi-Task" by 1.99 pp. PER. For the phoneme classifier, this indicates a reduction of negative transfer effects. However, the PER is still 2.46 pp. higher across the unseen languages compared to the baseline. Furthermore, the AER is 0.63 pp. higher than in "Multi-Task." This is consistent with Common Voice, where the AER of "Hierarchy" is also higher than of "Multi-Task."

Compared to "Hierarchy," our proposed blank removal method reduces PERs further by 5.15 pp. Therefore, it also outperforms the baseline by 2.69 pp. PER. Moreover, the AER is reduced by 3.54 pp. compared to "Multi-Task," making it our best performing model overall. "Hierarchy No Blanks" also still has the highest inter-language PER variance out of the evaluated models but by a less substantial

Table 2 Average phoneme and attribute error rates, their variances between languages, and phoneme insertion, deletion, and substitution rates for UCLA

Name	PER	PER σ^2	Insertions	Deletions	Substitutions	AER	AER σ^2
Phoneme Only	64.21%	154.37	3.76%	20.52%	39.93%	–	–
Multi-task	68.66%	133.46	**1.98%**	37.90%	**28.78%**	32.47%	61.25
Hierarchy	66.67%	164.86	3.25%	30.76%	32.66%	33.10%	78.59
Hierarchy No Blanks	**61.52%**	179.71	4.57%	**20.35%**	36.60%	**28.93%**	59.21

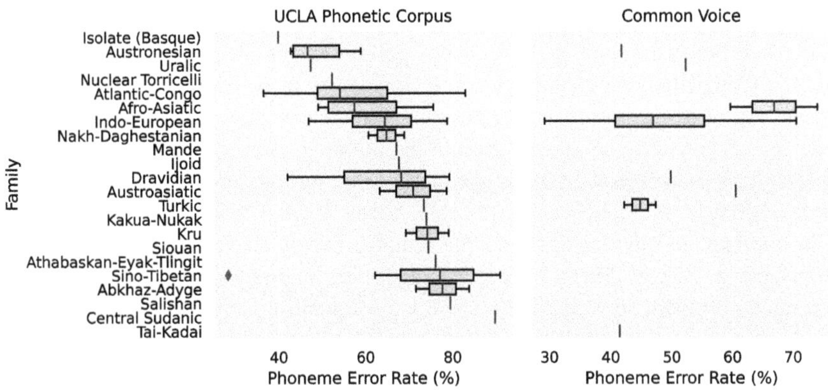

Fig. 3 PER distributions grouped by language families for cross-lingual transfer on UCLA and for the seen languages from Common Voice for reference. Languages were assigned to families based on Glottolog [7]. For language families where only one language occurs in the corpus, only the median line is shown

margin compared to the difference in variances observed in the Common Voice results.

When analyzing error types in more detail, the characteristics of the models differ in some aspects on the unseen languages compared to Common Voice. Deletion rates are still substantially higher in "Multi-Task" and "Hierarchy" compared to the baseline by 17.38 pp. and 10.24 pp., respectively, with substitution rates being lower than the baseline. However, "Hierarchy No Blanks" only reduces the deletion rate by 0.17 pp., with the increase in insertions by 0.81 pp. being comparatively larger. Instead, PER improvements are primarily caused by the reduction of substitutions by 3.33 pp. from the baseline. This suggests that there are two advantages of the blank removal method over the other multi-task architectures for cross-lingual phoneme recognition. Firstly, it is less susceptible to deletions, which can be observed in both corpora. Secondly, it better models the articulatory attributes seen in the AER reduction on UCLA. As a result, models trained with our proposed hierarchical multi-task architecture with blank removal can generalize better to unseen languages than our other multi-task models.

The cross-lingual transfer results of the "Hierarchy No Blanks" model are further divided into language families in Fig. 3 based on Glottolog [7]. Grouped supervised results from Common Voice are included for reference. In total, languages from eight language families occur in Common Voice. UCLA contains languages from a total of 21 languages families, of which one-third also occur in Common Voice. The Tai-Kadai family, which is represented by Thai in Common Voice, is the only language family that occurs in the training data but not UCLA.

We find that "Hierarchy No Blanks" transfers best to Basque. This could be explained by the model performing very well on Spanish. Although the two languages are not related, there are some similarities in their phonetics and the vast majority of Basque speakers are bilingual in Spanish or French [8]. After Basque, the model transfers best to the nine Austronesian languages in the corpus, despite Indonesian being the sole language from this family in the training data. PERs of Atlantic-Congo languages are the fourth lowest on average despite no languages of this family being present in the training data. The model also transfers slightly better to Afro-Asiatic languages than Indo-European languages even though there are only two Afro-Asiatic languages in the training data as opposed to 15 Indo-European ones.

In contrast, the model transfers particularly poorly to the two Abkhaz-Adyge languages in UCLA. Furthermore, utterances in languages originating from the American continent such as the Salishan and Siouan families from North America are also among those with the worst recognition performance. Some outliers with particularly high PER might also be caused by the noisy recording conditions of utterances [15]. An analysis of the effect of techniques for noise robustness in the context of cross-lingual transfer can be carried out in future work.

4.3 Combined Analysis

Models are further compared by analyzing the correlation between their PERs across languages. Correlations between every model pair in both corpora can be seen in Fig. 4. Overall, the PERs correlate strongly between models across languages with correlation coefficients being above 0.9 for every pair. Overall, correlations are stronger on Common Voice than UCLA. In particular, the "Phoneme Only," "Multi-Task," and "Hierarchy" models, for which the average PER is also very similar, correlate with $r^2 \geq 0.979$. In contrast, "Hierarchy No Blanks" has a slightly weaker correlation with the other models. Despite having a very similar architecture to

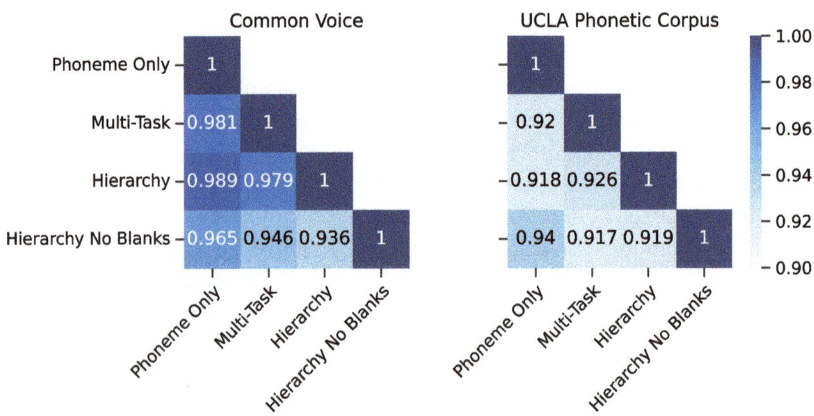

Fig. 4 Correlations of the PERs for each language between all models on Common Voice and UCLA, with the correlation coefficients displayed in each square

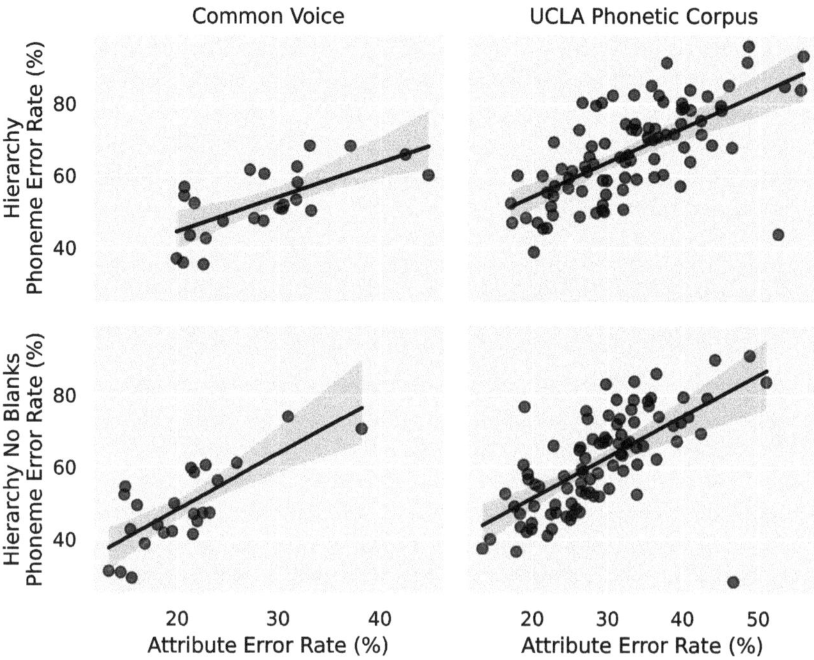

Fig. 5 Linear regression between Phoneme and Attribute Error Rates of "Hierarchy" (H) and "Hierarchy No Blanks" (HNB) on the Common Voice Test Set (H $r^2 = 0.457$; HNB $r^2 = 0.58$) and UCLA (H $r^2 = 0.434$; HNB $r^2 = 0.426$). The shaded areas represent the 95% confidence interval of each regression line estimated using bootstrap

"Hierarchy," the models correlate the least ($r^2 = 0.936$). Instead, it has the strongest correlation with the baseline $r^2 = 0.965$. On UCLA, the correlations are generally lower with $r^2 \in [0.917, 0.94]$. As in Common Voice, "Hierarchy No Blanks" has the strongest correlation with the baseline ($r^2 = 0.94$). Contrary to Common Voice, however, this correlation is the strongest overall on UCLA by a slight margin.

To get a more complete picture of the relation between attribute classification and phoneme recognition capabilities of the hierarchical multi-task models, linear regression plots between AER and PER in both test sets are shown in Fig. 5. On Common Voice, the moderate correlation between PER And AER from "Hierarchy No Blanks" ($r^2 = 0.58$) is much stronger than the correlation between the two metrics in the results of the regular "Hierarchy" model ($r^2 = 0.457$). This could imply that the improvements on an attribute level in the seen languages of the new architecture can be more effectively utilized by the phoneme recognizer than in "Hierarchy." More specifically, the higher correlation together with the overall lower PERs in most languages suggests that the improved attribute modeling can be better utilized by the phoneme classifier. This seems to particularly be the case in languages, in which attributes are also better recognized.

On UCLA, however, correlations between AER and PER are almost identical between "Hierarchy No Blanks" ($r^2 = 0.426$) and "Hierarchy" ($r^2 = 0.434$). This suggests that the benefits of "Hierarchy No Blanks" to phoneme classifications tend to apply more uniformly across languages instead of being related to the improved attribute modeling in a specific subset of languages. It could also be interpreted as the model's reliance on attribute-level classification performance remaining roughly the same as in "Hierarchy."

4.4 Phoneme Errors

In addition to PER and overall insertion, deletion, and substitution statistics, we also investigated common errors of these types produced by the "Hierarchical No Blank" model. The three most common errors of each type for Common Voice can be seen in Table 3. Most insertion errors in this corpus are erroneously recognized vowels, with [a] alone accounting for 11.6% of all insertion errors. This pattern can also be seen to a lesser extent in deletion errors, where [a] is the most deleted phoneme just above [i]. The model appears to also perform poorly on the recognition of nasal consonants. In particular [n] is the third most deleted phoneme, and the bilabial [m] and alveolar [n] nasals were often substituted with each other, making it the third most common substitution error. The two most common substitution errors are between two pairs of front vowels. The pairs are the close-mid ([e]) and open ([a]) front vowels and the close ([i]) and open-mid ([ɛ]) vowels.

When investigating deletion errors on a language level, we also found that some deletions coincide with common elision phenomena in these languages. For instance, 25.7% of schwas in German transcriptions are deleted. While it does not conclusively show that these deletions were primarily caused by this phenomenon, schwas are by far the most commonly elided vowels in the language in read speech [10]. This reveals another limitation in evaluation on phoneme transcriptions that were generated with a grapheme-to-phoneme system. Even with perfect accuracy, phoneme transcriptions from standard orthography may differ from the realizations by different speakers. Since transcriptions in UCLA are already phonemic, cross-lingual transfer results are unaffected.

The three most common errors of each type for UCLA are listed in Table 4. Similarly to Common Voice, vowels are the most deleted phonemes, with the

Table 3 The three most common errors for each error category and their occurrence frequency within their category on seen languages of the Common Voice test set

Insertion		Deletion		Substitution	
Phoneme	Frequency	Phoneme	Frequency	Phoneme Pair	Frequency
[a]	11.6%	[a]	6.1%	[e] ↔ [a]	2.4%
[i]	6.9%	[i]	5.4%	[i] ↔ [ɛ]	2.0%
[e]	6.5%	[n]	5.3%	[m] ↔ [n]	1.8%

Table 4 The three most common errors for each error category and their frequency within their category on unseen languages from UCLA

Insertion		Deletion		Substitution	
Phoneme	Frequency	Phoneme	Frequency	Phoneme Pair	Frequency
[k]	7.6%	[a]	11.1%	[aː] ↔ [a]	1.7%
[a]	5.5%	[i]	7.3%	[ɛ] ↔ [e]	1.5%
[n]	5.4%	[u]	5.5%	[u] ↔ [o]	1.3%

two most commonly deleted vowels being the same across corpora. However, the frequency of the individual vowel deletions is substantially higher, with particularly [a] deletions being 82% more common than on Common Voice. Furthermore, [a] insertions are frequent, like in Common Voice. In contrast, we find more insertions of consonants, with the high frequency of [n] insertions further showing the difficulties of "Hierarchy No Blanks" to recognize nasal consonants. The most commonly inserted phoneme in the corpus is [k]. Due to the overall rarity of insertion errors, a large fraction of these insertions come from Adyghe alone. After inspecting the phoneme inventory mapping for Adyghe, we noticed that one of the phonemes mapped from the acoustic model to [k] is [ŋ], where both are velar consonants but [ŋ] is voiced while [k] is unvoiced. This suggests that the [k] insertions are an indirect consequence of the high number of insertions of nasal consonants by our model. Substitution errors occurred frequently between the long [aː] and short [a] and the open-mid and close-mid front vowels, which are closely together in vowel space. However, in contrast to Common Voice we also find a common substitution error between the back vowels [u] and [o].

5 Conclusion

A novel hierarchical multi-task architecture is presented and evaluated together with a hybrid convolution-transformer acoustic model for phoneme recognition. In contrast to regular multi-task learning, the phoneme classifier receives articulatory attribute probabilities as additional inputs on each frame. Furthermore, we introduce a modification to the hierarchical architecture, where the logits of CTC blanks are removed from the outputs of each attribute classifier before computing and passing attribute probability distributions to the phoneme classifier. In addition to the evaluation on seen languages, we tackle cross-lingual transfer to low-resource languages. For zero-shot recognition in such languages, only their phoneme inventory is required.

On the seen languages in Common Voice, neither regular multi-task learning nor unmodified hierarchical multi-task learning outperforms our baseline without attribute classifiers. However, with the proposed blank removal method, hierarchical multi-task learning outperforms the baseline by 3.16 pp. PER. Furthermore, it reduces AER substantially by 5.97 pp. compared to regular multi-task learning.

When evaluating cross-lingual transfer to 95 languages from the UCLA Phonetic Corpus, we observe negative transfer effects both when using regular and, to a lesser degree, unmodified hierarchical multi-task learning. With our blank removal approach, we achieved a reduction of 2.69 pp. PER over the baseline and of 3.54 pp. AER over regular multi-task learning.

Future work may investigate the effect of grapheme-to-phoneme models across languages and their impact on cross-lingual generalization to unseen languages in more detail. Furthermore, future research may include the recognition of tones. This would allow for more complete and accurate transcriptions for low-resource tonal languages.

References

1. Ardila, R., Branson, M., Davis, K., Henretty, M., Kohler, M., Meyer, J., Morais, R., Saunders, L., Tyers, F.M., Weber, G.: Common voice: a massively-multilingual speech corpus. arXiv preprint arXiv:1912.06670 (2019)
2. Borleffs, E., Maassen, B.A., Lyytinen, H., Zwarts, F.: Measuring orthographic transparency and morphological-syllabic complexity in alphabetic orthographies: a narrative review. Read. Writ. **30**, 1617–1638 (2017)
3. Crawshaw, M.: Multi-task learning with deep neural networks: a survey. CoRR abs/2009.09796 (2020). https://doi.org/10.48550/ARXIV.2009.09796. https://arxiv.org/abs/2009.09796
4. Dauphin, Y.N., Fan, A., Auli, M., Grangier, D.: Language modeling with gated convolutional networks. In: International Conference on Machine Learning, pp. 933–941. PMLR (2017)
5. Glocker, K., Georges, M.: Hierarchical multi-task transformers for crosslingual low resource phoneme recognition. In: Proceedings of the 5th International Conference on Natural Language and Speech Processing (ICNLSP 2022), pp. 187–192. Association for Computational Linguistics, Trento (2022). https://aclanthology.org/2022.icnlsp-1.21
6. Graves, A., Fernández, S., Gomez, F.J., Schmidhuber, J.: Connectionist temporal classification: labelling unsegmented sequence data with recurrent neural networks. In: Proceedings of the 23rd International Conference on Machine Learning (2006). https://doi.org/10.1145/1143844.1143891
7. Hammarström, H., Forkel, R., Haspelmath, M., Bank, S.: glottolog/glottolog: Glottolog database 4.6 (2022). https://doi.org/10.5281/zenodo.6578297
8. Hualde, J.I.: Basque Phonology, Routledge, New York, NY, USA (2004)
9. Kingma, D.P., Ba, J.: Adam: a method for stochastic optimization. CoRR abs/1412.6980 (2015). https://doi.org/10.48550/ARXIV.1412.6980. https://arxiv.org/abs/1412.6980
10. Kohler, K.J., Rodgers, J.: Schwa deletion in German read and spontaneous speech. Spontaneous German Speech: Symb. Struct. Gestural Dyn. **35**, 97–123 (2001)
11. Lee, Y.T., Chen, X.B., Lee, H.S., Jang, J.S.R., Wang, H.M.: Multi-task learning for acoustic modeling using articulatory attributes. In: 2019 Asia-Pacific Signal and Information Processing Association Annual Summit and Conference (APSIPA ASC), pp. 855–861 (2019). https://doi.org/10.1109/APSIPAASC47483.2019.9023180
12. Li, X., Dalmia, S., Mortensen, D.R., Li, J., Black, A.W., Metze, F.: Towards zero-shot learning for automatic phonemic transcription. In: AAAI (2020). https://doi.org/10.1609/aaai.v34i05.6341
13. Li, X., Li, J., Metze, F., Black, A.W.: Hierarchical phone recognition with compositional phonetics. In: Proceedings of Interspeech 2021, pp. 2461–2465 (2021). https://doi.org/10.21437/Interspeech.2021-1803
14. Li, X., Metze, F., Mortensen, D.R., Black, A.W., Watanabe, S.: ASR2K: speech recognition for around 2000 languages without audio. In: Proceedings of Interspeech 2022, pp. 4885–4889 (2022). https://doi.org/10.21437/Interspeech.2022-10712

15. Li, X., Mortensen, D.R., Metze, F., Black, A.W.: Multilingual phonetic dataset for low resource speech recognition. In: ICASSP 2021–2021 IEEE International Conference on Acoustics, Speech and Signal Processing (ICASSP), pp. 6958–6962 (2021). https://doi.org/10.1109/ICASSP39728.2021.9413720
16. Moran, S., McCloy, D. (eds.): PHOIBLE 2.0. Max Planck Institute for the Science of Human History, Jena (2019). https://phoible.org/
17. Mortensen, D.R., Dalmia, S., Littell, P.: Epitran: precision G2P for many languages. In: Proceedings of the Eleventh International Conference on Language Resources and Evaluation (LREC 2018). European Language Resources Association (ELRA), Miyazaki (2018). https://aclanthology.org/L18-1429
18. Mortensen, D.R., Littell, P., Bharadwaj, A., Goyal, K., Dyer, C., Levin, L.S.: Panphon: a resource for mapping IPA segments to articulatory feature vectors. In: Proceedings of COLING 2016, the 26th International Conference on Computational Linguistics: Technical Papers, pp. 3475–3484. ACL (2016). https://aclanthology.org/C16-1328
19. Paszke, A., Gross, S., Massa, F., Lerer, A., Bradbury, J., Chanan, G., Killeen, T., Lin, Z., Gimelshein, N., Antiga, L., Desmaison, A., Kopf, A., Yang, E., DeVito, Z., Raison, M., Tejani, A., Chilamkurthy, S., Steiner, B., Fang, L., Bai, J., Chintala, S.: PyTorch: an imperative style, high-performance deep learning library. In: Wallach, H., Larochelle, H., Beygelzimer, A., d'Alché-Buc, F., Fox, E., Garnett, R. (eds.) Advances in Neural Information Processing Systems, vol. 32, pp. 8024–8035. Curran Associates, Inc. (2019). http://papers.neurips.cc/paper/9015-pytorch-an-imperative-style-high-performance-deep-learning-library.pdf
20. Qi, P., Zhang, Y., Zhang, Y., Bolton, J., Manning, C.D.: Stanza: a Python natural language processing toolkit for many human languages. In: Proceedings of the 58th Annual Meeting of the Association for Computational Linguistics: System Demonstrations (2020). https://doi.org/10.18653/v1/2020.acl-demos.14. https://aclanthology.org/2020.acl-demos.14
21. Siminyu, K., Li, X., Anastasopoulos, A., Mortensen, D.R., Marlo, M.R., Neubig, G.: Phoneme recognition through fine tuning of phonetic representations: a case study on Luhya language varieties. In: Proceedings of Interspeech 2021, pp. 271–275 (2021). https://doi.org/10.21437/Interspeech.2021-1434
22. Synnaeve, G., Xu, Q., Kahn, J., Grave, E., Likhomanenko, T., Pratap, V., Sriram, A., Liptchinsky, V., Collobert, R.: End-to-end ASR: from supervised to semi-supervised learning with modern architectures. ArXiv abs/1911.08460 (2019). https://doi.org/10.48550/ARXIV.1911.08460. https://arxiv.org/abs/1911.08460
23. Vaswani, A., Shazeer, N.M., Parmar, N., Uszkoreit, J., Jones, L., Gomez, A.N., Kaiser, L., Polosukhin, I.: Attention is all you need. ArXiv abs/1706.03762 (2017). https://doi.org/10.48550/ARXIV.1706.03762
24. Xiong, R., Yang, Y., He, D., Zheng, K., Zheng, S., Xing, C., Zhang, H., Lan, Y., Wang, L., Liu, T.: On layer normalization in the transformer architecture. In: Daumé III, H., Singh, A. (eds.) Proceedings of the 37th International Conference on Machine Learning. Proceedings of Machine Learning Research, vol. 119, pp. 10524–10533. PMLR (2020). https://proceedings.mlr.press/v119/xiong20b.html
25. Xu, Q., Baevski, A., Auli, M.: Simple and effective zero-shot cross-lingual phoneme recognition. ArXiv abs/2109.11680 (2021). https://doi.org/10.48550/ARXIV.2109.11680. https://arxiv.org/abs/2109.11680
26. Yang, Y.Y., Hira, M., Ni, Z., Chourdia, A., Astafurov, A., Chen, C., Yeh, C.F., Puhrsch, C., Pollack, D., Genzel, D., Greenberg, D., Yang, E.Z., Lian, J., Mahadeokar, J., Hwang, J., Chen, J., Goldsborough, P., Roy, P., Narenthiran, S., Watanabe, S., Chintala, S., Quenneville-Bélair, V., Shi, Y.: Torchaudio: building blocks for audio and speech processing. arXiv preprint arXiv:2110.15018 (2021). https://doi.org/10.48550/ARXIV.2110.15018. https://arxiv.org/abs/2110.15018
27. Zhu, C., An, K., Zheng, H., Ou, Z.: Multilingual and crosslingual speech recognition using phonological-vector based phone embeddings. In: 2021 IEEE Automatic Speech Recognition and Understanding Workshop (ASRU), pp. 1034–1041 (2021). https://doi.org/10.1109/ASRU51503.2021.9687966

Comparison of Error Correction and Extraction Approaches

Stefan Constantin and Alex Waibel

1 Introduction

Error corrections are crucial parts of dialogs because errors and ambiguities are difficult to avoid. For instance, imagine a household robot receiving the instruction "Put the cleaned spoons into the cutlery drawer." However, the robot is unaware of which specific drawer is designated as the cutlery drawer. It selects one of the drawers and places the spoons inside. If its choice is incorrect, the user must correct the robot by stating, for example, "No, into the drawer right of the sink." Alternatively, the robot can inquire about the location of the cutlery drawer by asking the user directly. The user's response, such as "It's the drawer right of the sink," serves as a correction since it resolves the ambiguity.

Another form of correction occurs when users change their initial instruction, saying something like "I changed my mind, the forks" or when the system misinterprets the user's request due to errors in automatic speech recognition or natural language understanding.

All these types of corrections can be handled in a uniform manner. As a result, we propose a component that takes a request and a correction as input and produces a corrected request as output. The correction replaces the corresponding phrases in the original request. This chapter focuses only on replacing entity phrases such as "drawer right of the sink", while other types of phrases, such as verb phrases, are beyond the scope of this study.

S. Constantin (✉) · A. Waibel
Institute for Anthropomatics and Robotics, Karlsruhe Institute of Technology, Karlsruhe, Germany
e-mail: stefan.constantin@kit.edu; waibel@kit.edu

M. Abbas (ed.), *Practical Solutions for Diverse Real-World NLP Applications*, Signals and Communication Technology,
https://doi.org/10.1007/978-3-031-44260-5_5

For instance, the request "Put the cleaned spoons into the cutlery drawer" along with the correction "No, into the drawer right of the sink" would be transformed into "Put the cleaned spoons into the drawer right of the sink".

Such a correction component offers two advantages over handling corrections directly within the main dialog component. First, it reduces the amount of training data required for the dialog component since corrections do not need to be learned if an open-domain correction component is available. Second, this type of correction component can be expanded to output pairs of reparandum and repair entities. In our example, the pair would be "cutlery drawer" and "drawer right of the sink". These entity pairs can be utilized, for instance, in a lifelong learning component of a dialog system to diminish the need for future corrections. Consequently, the robot can learn which specific drawer corresponds to the cutlery drawer.

2 Prior Research

This work is an extension of work originally presented in [4]. Previous research has been carried out in the field of interactive repair dialog. In one study [22], a multimodal approach was employed where users could indicate incorrect phrases and correct them through respeaking, spelling, selecting alternatives from the n-best list of the automatic speech recognition component, or using handwriting. The strategies for error correction were further enhanced in another study [25] by considering the context. Additional evaluations of these approaches were conducted in a dictation system with real users in subsequent studies [23, 24]. An exploration of different human strategies for error correction was presented in a study by [9].

Sagawa et al. [20] proposed an error handling component based on correction grammars, which offered the advantage of domain-independent use. However, it required a grammar-based dialog system. Griol and Molina [10] proposed an error correction detection module and strategies to handle detected errors. The corrected request then had to be processed by the Spoken Language Understanding component, which necessitated adapting the component to accommodate possible corrections for each domain. Kraljevski and Hirschfeld [13] proposed a domain-independent correction detection method by examining hyperarticulation in speech, without utilizing features other than hyperarticulation.

In another study [1], a system was introduced that detected errors in automatic speech recognition transcripts and prompted users for corrections.

Some studies have also explored automatic error correction without user interaction. For instance, [31] employed a character-based approach to correct language errors and mitigate issues arising from orthographic errors. Wen et al. [29] utilized a multi-task setup to correct automatic speech recognition outputs and perform natural language understanding.

The task of request correction discussed in the introduction is closely related to disfluency removal. Disfluency removal involves identifying the reparandum (the entity to be replaced), the interruption point (where the correction starts), the

spoon into the drawer uh sink

reparandum interruption pt. interregnum repair

Fig. 1 Disfluent utterance annotated with repair terminology [4]

spoons into the drawer uh sink
C C C D D C

Fig. 2 Disfluent utterance labeled with copy and delete labels [4]

spoon into the drawer no forks

reparandum interruption pt. interregnum repair

Fig. 3 Request and correction phrase annotated with repair terminology [4]

interregnum (the signal phrase for the correction), and the repair phrase (the correct entity) [21].

Figure 1 illustrates a disfluent utterance annotated with this terminology.

Extensive research has been carried out on disfluency removal [3, 8, 28, 11]. These studies share the assumption that removing the tokens from a disfluent utterance is sufficient to obtain a fluent utterance. Figure 2 illustrates a disfluent utterance annotated with copy and delete labels as depicted in these works.

Nevertheless, in the context of corrections, it is possible for long-distance replacements to occur. This implies that there are important words between the reparandum and the repair that should not be deleted. Figure 3 illustrates an example of a long-distance replacement in the context of corrections.

3 Benchmark Data

Our adapted dataset is derived from the natural language annotations of the EPIC-KITCHENS-100 dataset [5, 6]. The EPIC-KITCHENS-100 dataset consists of 100 hours of recordings capturing actions in a kitchen setting. Each action is annotated with a corresponding verb, entities, and their respective classes. For example, an annotation may be "put pizza slice into container" with the verb "put-into" and the entities "slice:pizza" and "container".

The annotations in this dataset typically include one verb and zero to six entities. The verb, its corresponding verb class, the entities, and their corresponding entity classes are explicitly saved for each annotation. The order of the entities and their classes aligns with the annotation. If the verb includes a preposition, it is saved along with the verb. The entities are represented in a hierarchical structure, with the most general term on the left and increasing specificity toward the right, separated by colons.

The training dataset of the EPIC-KITCHENS-100 dataset comprises 67,218 annotations, while the validation dataset contains 9669 annotations. There is no separate test dataset. Some annotations may appear multiple times due to different recordings that have some common annotations, resulting in 15,968 unique annotations in the training dataset and 3835 unique annotations in the validation dataset.

For our specific dataset, we only utilize annotations that involved one or two entities. Annotations without any entities are excluded, as we require at least one entity that could be corrected. Annotations with more than two entities accounted for less than 1.15% of the total annotations, so they were excluded by us for the purpose of dataset balancing.

The verb classes in the original EPIC-KITCHENS-100 dataset exhibited imbalances. To achieve better balance in the validation dataset, we removed annotations belonging to verb classes that occurred frequently. We aim for a more evenly distributed dataset to evaluate how well the model can handle diverse verb classes. We determined the desired number of remaining annotations for a verb class, denoted as r, by dividing the total number of annotations, denoted as a, by 100. However, we set a minimum number of remaining annotations: 2 for one-entity annotations ($r = max(2, a/100)$) and 4 for two-entity annotations ($r = max(4, a/100)$). In certain instances, the EPIC-KITCHENS-100 dataset lacks the desired number of remaining annotations for a particular verb class and then we used the possible number. We carefully selected minimal examples to create a dataset that is nearly balanced: we included 142 annotations with one entity and 122 annotations with two entities. When gathering annotations for a verb class, we ensured an equal distribution of verbs belonging to that class. The reduced validation dataset contains a total of 264 annotations.

In the EPIC-KITCHENS-100 dataset, the training and validation datasets share many similarities. All 78 verb classes occurring in the validation dataset also occur in the training dataset. Furthermore, out of the 372 first-level words in the entity hierarchies of the validation dataset, 346 of them are also found in the training dataset. To introduce more diversity between the training and validation datasets, we decided to make some reductions.

First, we removed the 49 least frequently occurring verb classes from the training dataset (in total 98 verb classes are in the training dataset). Additionally, we eliminated all entities from the training dataset if their first part (in the entity hierarchy) was already present in the validation dataset. This means that if an annotation in the validation dataset has, for example, "bowl:washing:up," any annotations in the training dataset containing "bowl:salad" were removed. After this reduction process, 4822 annotations remain in the training dataset.

To train and evaluate the error correction detection and correction components, we needed to add corrections to the annotations. For the training and validation datasets, we synthetically generated the corrections. There were three options for entity replacement: replacing the first entity, replacing the second entity, or replacing both entities. We uniformly selected which option to apply. In cases where both entities needed correction, we uniformly determined the order in which they should be corrected. We had eight templates for introducing the correction phrase followed by the corrected entities in the training dataset and six templates for the validation dataset. An entity could be replaced by another entity from an annotation of the same verb class, in the same position. Examples of corrected entities include phrases like "Be so kind and pick the oregano" for the request and "it's the chili" for the correction, or "Could you put the tin in the Cupboard?" for the request and "no the olives in the Fridge" for the correction.

For the test dataset, we enlisted the help of nine human data collectors who were given the freedom to write their own corrections. They were only informed about which entities should be replaced with what other entities, but they were allowed to use synonyms for the other entities. The corrections were categorized into three types: correction to a wrong action of the robot, clarification, or correction due to a change in the user's mind. These categories were equally distributed among the corrections.

To increase the natural language variety in the training and validation datasets, we added 19 and 14 templates before the narration, respectively. Additionally, we added the article "the" before the entities because it is missing in the original EPIC-KITCHENS-100 dataset. For the test dataset, we used the narrations from our validation dataset and asked the same nine annotators who created the corrections to paraphrase them.

The test dataset pose a greater challenge compared to the validation dataset as it diverge even further from the training dataset. The nine data collectors were instructed to use a wide range of natural language expressions.

Using the 4822 annotations from the reduced training dataset, combined with various data augmentations, we generated 52,357 request and correction pairs for the error correction training dataset. The error correction validation dataset consists of 264 request and correction pairs, while the error correction test dataset contains 960 request and correction pairs.

To train and evaluate the error correction detection, examples were needed where the last utterance was not a correction. To achieve this, the second last and last utterances were created by compiling all the requests from the error correction data. The requests were shuffled for the last utterance. This approach effectively doubled the number of examples compared to the correction examples. Therefore, the error correction detection and correction training dataset has 104,714 pairs, the error correction detection and correction validation dataset has 528 pairs, and the error correction detection and correction test dataset has 1920 pairs.

The target for the error correction datasets consists of the corrected request and the reparandum repair pairs. The target for the error correction detection and correction datasets depends on whether the source has a request and correction pair or a request and request pair. In the former case, where an error correction exists, the target is the same as in the error correction datasets. In the latter case, the target is to copy both requests.

There is also an additional dataset called the error correction detection dataset. The sources for this dataset are the same as in the error correction detection and correction dataset, but the target is a binary value indicating whether there is a correction or not.

The described datasets are created in different forms to cater to various approaches. For the sequence labeling approach, the source tokens were labeled with different labels (Fig. 4). For the sequence-to-sequence approach with generative token generation, source and target pairs were created (Fig. 5). For the sequence-to-sequence approach with generation by copying source tokens, the order of copy operations was added. Additionally, separator tokens required in the target were inserted into the source (Fig. 6).

For the few shot approaches with the large language model GPT-3 [2] prompts are needed that introduce the task of error correction detection and/or error correction to the model. We have two approaches for generating the prompts: First, a static prompt generation that generates one prompt that is the same for all evaluation examples, and then we have a dynamic prompt generation that generates a specific prompt for every example that has to be evaluated [14]. In the first approach, for the error correction detection and error correction dataset, we found the two examples out of the validation dataset that are most similar to all of the examples in the training dataset where the last utterance is an error correction and two examples

```
would C
it C
be C
possible C
to C
wash C
the C
table R1
? C
| D
no D
the D
wok S1
instead D
of D
the D
table D
. D
```

Fig. 4 Sequence labeling data example [4]

```
source file: Would it be possible to wash the table ? | no the Wok instead of the table .
target file: Would it be possible to wash the Wok ? | table -> Wok
```

Fig. 5 Sequence to sequence with fixed vocabulary data example [4]

```
source file: Would it be possible to wash the table ? | no the Wok instead of the table . - ->
target file: Would it be possible to wash the Wok ? | table -> Wok
copy target file (considering the T5 prefix and the T5 tokenization): 3 4 5 6 7 8 9 16 17 11 12 13
10 26 27 16 17 28
```

Fig. 6 Sequence to sequence with copy source token approach data example [4]

utterance 1: could you please empty the bowl contents into the saucepan
utterance 2: i meant into the cup .
output: could you please empty the bowl contents into the cup | saucepan -> cup

utterance 1: an you please empty the bowl contents into the saucepan .
utterance 2: could you be so kind and take the garlic press .
output: can you please empty the bowl contents into the saucepan . | could you be so kind and take the garlic press . |

utterance 1: Please empty the bowl contents into the saucepan .
utterance 2: It's into the can .
output: Please empty the bowl contents into the can . | saucepan -> can

utterance 1: can you please turn using the Tongs .
utterance 2: could you please empty the bowl contents into the saucepan .
output: can you please turn using the Tongs . | could you please empty the bowl contents into the saucepan . |

Fig. 7 Static prompt of the error correction detection and error correction dataset [4]

where the last utterance is no error correction by using Sentence-BERT [19]. To get the two most similar examples where the last utterance is an error correction, we calculated to all correction examples of the validation dataset the similarity to all examples where the last utterance is a correction in the training dataset and averaged all the similarity scores of one validation example. We chose the two validation examples with the highest averaged similarity score as our most similar examples. The same procedure was applied to all examples where the last utterance is no error correction. In Fig. 7 the static prompt for the error correction detection and error correction dataset is depicted. For the error correction dataset, the prompt contains only the two most similar examples where the last utterance is a correction. In the second prompt generation approach, we searched for every example that should be evaluated the two most similar examples in the training dataset where the last utterance is a correction and in case of the datasets where error correction detection is included additionally the two most similar examples in the training dataset where the last utterance is no correction and we generated with these similar examples a prompt comparable to the prompts of the static prompt generation approach. The error correction detection dataset is similar to the error correction detection and error correction dataset except that the output is the binary value whether the last utterance is a correction or not. For the prompt generation approaches, we used the token "correction" for an example where the last utterance is a correction and "no correction" for the examples where the last utterance is no correction.

The GPT-3 model can also be fine-tuned. As training dataset for fine-tuning, we randomly chose 1000 examples where the last utterance is a correction and 1000 examples where the last utterance is no correction. For the error correction dataset, we used in the fine-tuning dataset the token "correction" for an example where the last utterance is a correction and "okay" for the examples where the last utterance is no correction to utilize the fine-tuned GPT-3 model as classification model that outputs only one token.

4 Approaches

We developed six different methods for error correction and extraction. The first approach labels sequences, while the second approach generates sequences where the output tokens are selected from a predefined vocabulary. The third approach also generates sequences, but this time the output tokens are copied from the source tokens. The remaining three approaches are based on the GPT-3 model. Among these three approaches, one uses a fixed prompt for all test examples, another employs a dynamic prompt based on the test example, and the third approach does not use a prompt but is fine-tuned.

In the sequence labeling approach, each word is assigned one of several labels: C (copy), D (delete), R1 (entity 1, possibly to be replaced), R2 (entity 2, possibly to be replaced), S1 (entity to replace entity 1), or S2 (entity to replace entity 2). For the correction target, the entities labeled as S1 and S2 are used to replace the entities labeled as R1 and R2, respectively. Regarding the extraction target, the output consists of pairs: R1 and S1, as well as R2 and S2 if a replacement is available for the first or second entity, respectively. Figure 8 provides an example where a request and its corresponding correction pair are labeled, and both correction and extraction targets are provided.

To implement the sequence labeling approach, we suggest fine-tuning the cased BERT large model, which consists of 24 Transformer encoder blocks, a hidden size of 1024, 16 self-attention heads, and 340 million parameters [7].

For the sequence-to-sequence approach, where the output tokens are selected from a fixed vocabulary, we propose fine-tuning a T5 large model [18]. The T5 model is a pre-trained Transformer network [26], and the T5 large model has the following specifications: 24 Transformer encoder blocks, 24 Transformer decoder blocks, a hidden size of 1024 for input and output, a hidden size of 4096 for inner layers, 16 self-attention heads, and 737 million parameters.

To calculate the probability distribution over the fixed vocabulary V, the following procedure can be employed:

$$P_{\text{generate}}(V) = softmax(dec^T \cdot W_{\text{generate}}),$$

where dec is the output of the Transformer decoder and $W_{\text{generate}} \in \mathbb{R}^{\text{hidden size decoder} \times \text{vocabulary size}}$ is a during training adjustable matrix.

We refer to this specific T5 model as T5 generate.

request || correction	put the milk into the shelf	no the soy milk into the left shelf
labels	C R1 R1 R2 R2 R2	D S1 S1 S1 S2 S2 S2 S2
corrected request	put the soy milk into the left shelf	
pairs of reparandum and repair entity	milk → soy milk - into the shelf → into the left shelf	

Fig. 8 Error correction example [4]

In the corrected request, the contained tokens only belong to the input sequence. To leverage this characteristic, we have devised a pointer network model [27] using the T5 large model. This model determines which input token has the highest likelihood of being copied to the output sequence. This constitutes our third approach. The probability distribution over the tokens of the input sequence V can be computed as follows:

$$P_{\text{copy}}(V') = softmax(dec^T \cdot enc^T),$$

where dec is the output of the Transformer decoder and $enc \in \mathbb{R}^{\text{source input length} \times \text{embedding size}}$ is the output of the Transformer encoder.

To leverage the knowledge encoded in the pre-trained model, instead of using the position of the source input token, we feed the encoder with the source input token that has the highest probability. In the generation stage, the copy mechanism is only utilized, making it similar to a regular T5 model. In order to include the separators in the output, we append them to the source input so that they can also be copied. We refer to this modified T5 model as T5 copy.

To determine whether an utterance serves as a correction for the previous request command, the three approaches described above can also be employed. In the sequence labeling approach, if all output labels are C, no error correction is detected; otherwise, an error correction is present. The sequence-to-sequence approaches detect an error correction if the source and target (without separators) are not identical; otherwise, there is no error correction. In the T5 copy approach, the original source is used for the comparison, rather than the source with inserted separators.

In addition to these three approaches, sequence classification can also be used for error correction detection. For this, we propose fine-tuning the cased BERT large model, which consists of 24 Transformer encoder blocks, a hidden size of 1024, 16 self-attention heads, and 340 million parameters [7].

In addition to the previous approaches and to avoid fine-tuning, a large language model can be used as few shot model. We used the 175 billion parameter GPT-3 [2], more specific the text-davinci-003 model, as large language model. To be able to do the error correction detection and/or error correction task, one (one shot) or more (few shot) examples must be given as prefix in the prompt. We used two examples where the last utterance is an error correction and two examples where the last utterance is no error correction in addition for the tasks including error detection. The difference between static prompt prefix and dynamic prompt prefix is explained in Sect. 3. As comparison to the few shot approaches with static and dynamic prompts, we fine-tune the mentioned 175 billion parameter GPT-3 model.

5 Experimental Setup

For Sentence-BERT, we used the original paper implementation [19], and for GPT-3, we used the officially Python API [16].

For the other models discussed in Sect. 4, we employed the HuggingFace [30] PyTorch [17] BERT and T5 models and published our implementations and models for these other models.[1]

6 Results

In this section, we will begin by evaluating different approaches for error correction detection, as described in Sect. 4. Following that, we will evaluate the error correction component approaches outlined in the same section. Third, we will compare the effectiveness of a pipeline approach, where error correction detection and error correction are separate components, with an end-to-end approach. All evaluations were conducted using the datasets described in Sect. 3.

The sequence classification and labeling approaches were fine-tuned for one epoch using the following hyperparameters: AdamW optimizer [15] with a learning rate of $2 \cdot 10^{-5}$, a batch size of 32, and a maximum input length of 128.

The T5 generate and T5 copy models were fine-tuned for one epoch with the following hyperparameters: Adam optimizer [12] with a learning rate of $2.5 \cdot 10^{-4}$, a batch size of 24, and a maximum input length of 128. In the embedding layer, the first two encoder blocks were frozen.

The GPT-3 model used with the generated static and dynamic prompts was not fine-tuned, while the fine-tuned GPT-3 models were fine-tuned for four epochs by OpenAI.

The results of the GPT-3 approaches were evaluated in a case-insensitive manner and without considering spaces. This was done because some of the data contain spaces between words and punctuation, or words start with an uppercase character, even when they are not supposed to. Non-GPT-3 approaches copy the grammatically incorrect elements, but the GPT-3 approaches correct them. In general, this is not a disadvantage and should not affect the evaluation.

The results of the error correction detection components are shown in Table 1. Accuracy measures the number of correctly classified examples, precision indicates the number of true positive examples out of the positively classified examples, recall measures how many of the positive examples are detected by the component, and the F_1-score is the harmonic mean of precision and recall. Precision, recall, and F_1-score were calculated for both detecting corrections as positive examples and detecting no corrections as positive examples to gain better insights into the

[1] https://github.com/msc42/seq2seq-transformer https://github.com/msc42/seq-labeling-and-classification

Table 1 Evaluation results of the error correction detection, all models except the classification were trained on the error correction detection and error correction dataset, and the classification was trained on the error correction detection dataset

			Detecting corrections			Detecting no corrections		
Dataset	Model	Accuracy	Precision	Recall	F_1-score	Precision	Recall	F_1-score
Valid	Classification	100%	100%	100%	100%	100%	100%	100%
Valid	Seq. labeling	100%	100%	100%	100%	100%	100%	100%
Valid	T5 generate	98.67%	98.13%	99.24%	98.68%	99.23%	98.11%	98.67%
Valid	T5 copy	71.78%	63.92%	100%	77.99%	100%	43.56%	60.69%
Valid	GPT-3 (static prompt)	78.60%	70.35%	98.86%	82.20%	98.09%	58.33%	73.16%
Valid	GPT-3 (dynamic prompt)	59.47%	55.43%	96.59%	70.44%	86.76%	22.35%	35.54%
Valid	GPT-3 fine-tuned for classification	100%	100%	100%	100%	100%	100%	100%
Test	Classification	87.86%	99.86%	75.83%	86.20%	80.52%	99.90%	89.17%
Test	Seq. labeling	88.49%	99.87%	77.08%	87.01%	81.34%	99.90%	89.67%
Test	T5 generate	84.01%	96.58%	70.52%	81.52%	76.78%	97.50%	85.91%
Test	T5 copy	77.45%	73.63%	85.52%	79.13%	82.73%	69.38%	75.47%
Test	GPT-3 (static prompt)	80.94%	73.50%	96.77%	83.54%	95.27%	65.10%	77.35%
Test	GPT-3 (dynamic prompt)	62.92%	57.82%	95.52%	72.03%	87.13%	30.31%	44.98%
Test	GPT-3 fine-tuned for classification	91.67%	99.75%	83.54%	90.93%	85.84%	99.79%	92.29%

performance of the differently trained models. The sequence classification, sequence labeling, and fine-tuned GPT-3 approaches achieved the highest accuracy on the validation dataset, all scoring 100%. On the test data, the fine-tuned GPT-3 approach performed the best with an accuracy of 91.67%, followed by the sequence labeling approach with 88.49% and the sequence classification approach with 87.86%. The three approaches with 100% validation accuracy showed a high recall for detecting examples where the last utterance contains no error correction (99.79% for the fine-tuned GPT-3 model and 99.90% for the other two approaches) and a high precision for examples where the last utterance contains no error correction (85.84% for the fine-tuned GPT-3 model, 81.34% for the sequence labeling approach, and 80.52% for the sequence classification approach). This means that if there is no correction, the component correctly identifies it in most cases and avoids unnecessary corrections. This is a desirable quality since it is better to not detect a correction than to correct something that is already correct. However, the results for detecting corrections, with a recall of 83.54%, 77.08%, and 75.83%, and a precision of 99.75%, 99.87%, and 99.86% on the test dataset for the fine-tuned GPT-3 model, sequence labeling, and sequence classification approaches, respectively, are also good. In some cases where the component fails, it is particularly challenging to detect the correction, such as in the example "Kindly turn off the heat on the oven

Table 2 Evaluation results of the error correction (metric accuracy), the end-to-end (E2E) models were trained on the error correction detection and error correction dataset and the other models were trained on the error correction dataset

	Validation dataset			Test dataset		
Model	Correction	Extraction	Both	Correction	Extraction	Both
Seq. labeling	96.21%	94.70%	94.70%	40.10%	48.75%	39.06%
E2E seq. labeling	96.59%	95.08%	95.08%	39.27%	43.54%	38.65%
T5 generate	92.80%	95.83%	91.29%	73.65%	77.81%	71.98%
E2E T5 generate	96.21%	95.08%	94.70%	37.40%	38.75%	36.25%
T5 copy	50.38%	87.12%	50.00%	50.52%	62.19%	47.92%
E2E T5 copy	70.83%	92.42%	68.94%	27.50%	35.00%	25.31%
GPT-3 (static prompt)	45.08%	65.15%	40.91%	53.02%	70.62%	49.58%
GPT-3 (dynamic prompt)	48.48%	74.24%	44.70%	45.63%	73.33%	42.71%
GPT-3 fine-tuned	91.29%	57.58%	55.68%	74.17%	69.48%	62.60%
E2E GPT-3 fine-tuned	59.47%	37.88%	34.47%	43.65%	40.52%	35.31%

| Please turn off the water tap on the oven." The T5 generate approach performed worse, achieving an accuracy of 98.67% on the validation dataset and 84.01% on the test dataset.

The results of the error correction components are presented in Table 2. The evaluation for error correction utilized the accuracy metric. The correction is considered correct if the predicted correction matches the reference correction. The extraction of reparandum and repair pairs is considered correct if the predicted pairs are equal to the reference pairs, ignoring the order and entities mapping to themselves. Both correction and extraction are deemed correct if both aspects are accurate. The error correction datasets were used for this evaluation. On the validation dataset, the sequence labeling approach trained on the error correction detection and error correction dataset achieved the highest overall accuracy of 95.08%. The accuracy for correction was 96.59%, and for extraction, it was 95.08%. On the test dataset, the T5 generate approach trained on the error correction dataset performed the best with an accuracy of 71.98% (correction accuracy: 73.65%, extraction accuracy: 77.81%). In general, all approaches trained on the error correction detection and error correction dataset exhibited higher accuracy on the validation dataset (except for the fine-tuned GPT-3 model), while all approaches trained on the error correction dataset achieved higher accuracy on the test dataset. The T5 copy extraction could be optimized by keeping track of the order of copy operations, stopping after finishing the correction, and using this information to reconstruct the reparandum and repair pairs. However, we decided not to pursue this optimization since the correction results are significantly worse, and further improvements would be minimal.

The results of the error correction detection and error correction components are presented in Table 3. The same accuracy metric was used as in the error correction evaluation. In the pipeline approach, the sequence labeling approach was employed

Table 3 Evaluation results of the error correction detection and error correction (metric accuracy), the end-to-end (E2E) models were trained on the error correction detection and error correction dataset and the other models were trained on the error correction dataset, "and" means that the error correction detection was done by the best error correction detection model (sequence labeling) and the error correction detection by the model mentioned after the "and" if a correction was detected

	Validation dataset			Test dataset		
Model(s)	Correction	Extraction	Both	Correction	Extraction	Both
Detection and seq. labeling	98.11%	97.35%	97.35%	67.08%	68.44%	66.72%
Detection and E2E seq. labeling	98.30%	97.54%	97.54%	69.58%	71.77%	69.27%
E2E seq. labeling	98.30%	97.54%	97.54%	69.58%	71.77%	69.27%
Detection and T5 generate	96.40%	98.67%	96.40%	78.49%	79.84%	77.81%
Detection and E2E T5 generate	98.11%	98.30%	98.11%	68.33%	68.80%	67.66%
E2E T5 generate	97.54%	97.16%	96.40%	68.07%	68.49%	66.88%
Detection and T5 copy	75.19%	94.13%	75.00%	68.44%	72.71%	66.93%
Detection and E2E T5 copy	85.42%	96.59%	84.66%	63.28%	66.77%	62.08%
E2E T5 copy	69.70%	78.03%	56.25%	55.00%	58.91%	47.40%
Detection and GPT-3 (static prompt)	72.54%	82.58%	70.45%	71.76%	77.86%	70.52%
Detection and E2E GPT-3 (static prompt)	65.34%	74.05%	63.83%	69.95%	72.92%	68.23%
E2E GPT-3 (static prompt)	62.31%	59.85%	48.48%	69.84%	67.34%	60.68%
Detection and GPT-3 (dynamic prompt)	74.24%	87.12%	72.35%	69.17%	79.06%	68.02%
Detection and E2E GPT-3 (dynamic prompt)	69.32%	81.25%	66.86%	69.58%	75.62%	67.29%
E2E GPT-3 (dynamic prompt)	59.09%	69.70%	51.33%	66.46%	70.83%	58.23%
Detection and GPT-3 fine-tuned	95.64%	78.79%	77.84%	83.08%	79.30%	75.42%
Detection and E2E GPT-3 Fine-tuned	100%	100%	100%	82.89%	78.51%	76.12%
E2E GPT-3 fine-tuned	96.59%	29.73%	29.36%	81.88%	57.86%	56.41%

for error correction detection, where an example is considered to have no correction if all labels are "C". After error correction detection, the error correction process is performed. We evaluated all three approaches described in Sect. 4 in their versions trained on the error correction detection and error correction dataset, as well as their versions trained on the error correction dataset. In the end-to-end setting, a single component handles both error correction detection and error correction in a single run. The pipeline approach with the T5 generate approach for error correction achieved the highest accuracy, with 96.40% on the validation dataset and 77.81% accuracy on the test dataset.

The evaluation results indicate that the test dataset presents a greater challenge compared to the validation dataset. The nine data collectors were able to introduce even more natural language variations than what is found in the validation dataset.

7 Conclusions and Further Work

There is a huge accuracy difference between the approaches. The T5 copy model fine-tuned with the error correction detection and error correction dataset is the worst approach in the real-world error correction detection and error correction test dataset with an accuracy of 47.70%. Whereas, the pipeline with the fine-tuned sequence labeling BERT model as error correction detection and the fine-tuned sequence-to-sequence T5 generate model as error correction is the best approach with an accuracy of 77.81% on that test dataset. This relatively good result on real data shows that this approach is learning the concept of corrections and thereby offers a utility component for a dialog system that does not need to be trained for error correction for every domain individually. Furthermore, the extraction of the reparandum and repair pairs additionally offers the possibility to learn from them in a lifelong learning component and with which the user will not have to give the same corrections every single time. The approach with the best result needed 104,714 training examples, but the fine-tuned end-to-end GPT-3 model used in the pipeline with the sequence labeling approach as error correction detection was only slightly worse with 76.12% accuracy and was only trained with 2000 training examples. In future research, better improved large language models could give better performance with less data needed and would make the creation of an error correction detection and error correction component in the future easier.

A further future research goal is to be able to correct all phrases and not only entity phrases and to use speech with the problems of word errors in the upstream component Automatic Speech Recognition.

Acknowledgments This work has been supported by the German Federal Ministry of Education and Research (BMBF) under the project OML (01IS18040A).

References

1. Béchet, F., Favre, B.: ASR error segment localization for spoken recovery strategy. In: 2013 IEEE International Conference on Acoustics, Speech and Signal Processing, pp. 6837–6841 (2013)
2. Brown, T., Mann, B., Ryder, N., Subbiah, M., Kaplan, J.D., Dhariwal, P., Neelakantan, A., Shyam, P., Sastry, G., Askell, A., Agarwal, S., Herbert-Voss, A., Krueger, G., Henighan, T., Child, R., Ramesh, A., Ziegler, D., Wu, J., Winter, C., Hesse, C., Chen, M., Sigler, E., Litwin, M., Gray, S., Chess, B., Clark, J., Berner, C., McCandlish, S., Radford, A., Sutskever, I., Amodei, D.: Language models are few-shot learners. In: Advances in Neural Information Processing Systems (2020)

3. Cho, E., Niehues, J., Waibel, A.: Machine translation of multi-party meetings: segmentation and disfluency removal strategies. In: Proceedings of the 11th International Workshop on Spoken Language Translation (IWSLT) (2014)
4. Constantin, S., Waibel, A.: Error correction and extraction in request dialogs. In: Proceedings of the 5th International Conference on Natural Language and Speech Processing (ICNLSP 2022) (2022)
5. Damen, D., Doughty, H., Farinella, G.M., Fidler, S., Furnari, A., Kazakos, E., Moltisanti, D., Munro, J., Perrett, T., Price, W., Wray, M.: The EPIC-KITCHENS dataset: collection, challenges and baselines. In: IEEE Transactions on Pattern Analysis and Machine Intelligence (TPAMI) (2020)
6. Damen, D., Doughty, H., Farinella, G.M., Furnari, A., Kazakos, E., Ma, J., Moltisanti, D., Munro, J., Perrett, T., Price, W., Wray, M.: Rescaling egocentric vision: collection, pipeline and challenges for EPIC-KITCHENS-100. Int. J. Comput. Vis. (IJCV) **130**(1), 33–55 (2022)
7. Devlin, J., Chang, M.-W., Lee, K., Toutanova, K.: BERT: pre-training of deep bidirectional transformers for language understanding. In: Proceedings of the 2019 Conference of the North American Chapter of the Association for Computational Linguistics (NACL) (2019)
8. Dong, Q., Wanga, F., Yang, Z., Xu, W.C.S., Xu, B.: Adapting translation models for transcript disfluency detection. In: Proceedings of the Thirty-Third Conference on Artificial Intelligence (AAAI) (2019)
9. Gieselmann, P.: Comparing error-handling strategies in human-human and human-robot dialogues. In: Proceedings of the 8th Conference on Natural Language Processing (Konferenz zur Verarbeitung natrlicher Sprache, KONVENS) (2006)
10. Griol, D., Molina, J.M.: A framework for improving error detection and correction in spoken dialog systems. Soft Comput. **20**, 4229–4241 (2016)
11. Jamshid Lou, P., Anderson, P., Johnson, M.: Disfluency detection using auto-correlational neural networks. In: Proceedings of the 2018 Conference on Empirical Methods in Natural Language Processing (EMNLP) (2018)
12. Kingma, D.P., Ba, J.: Adam: a method for stochastic optimization. In: 3rd International Conference on Learning Representations (ICLR), Conference Track Proceedings (2015)
13. Kraljevski, I., Hirschfeld, D.: Hyperarticulation of corrections in multilingual dialogue systems. In: Proceedings of the 18th Annual Meeting of the International Speech Communication Association (Interspeech) (2017)
14. Liu, J., Shen, D., Zhang, Y., Dolan, B., Carin, L., Chen, W.: What makes good in-context examples for GPT-3? In: Proceedings of Deep Learning Inside Out (DeeLIO 2022): The 3rd Workshop on Knowledge Extraction and Integration for Deep Learning Architectures, pp. 100–114, Dublin, and Online, May 2022. Association for Computational Linguistics
15. Loshchilov, I., Hutter, F.: Decoupled weight decay regularization. In: 7th International Conference on Learning Representations (ICLR) (2019)
16. OpenAI L.P.: OpenAI-Python. https://github.com/openai/openai-python (2023)
17. Paszke, A., Gross, S., Massa, F., Lerer, A., Bradbury, J., Chanan, G., Killeen, T., Lin, Z., Gimelshein, N., Antiga, L., Desmaison, A., Kopf, A., Yang, E., DeVito, Z., Raison, M., Tejani, A., Chilamkurthy, S., Steiner, B., Fang, L., Bai, J., Chintala, S.: PyTorch: an imperative style, high-performance deep learning library. In: Advances in Neural Information Processing Systems 32, pp. 8024–8035. Curran Associates, Inc. (2019)
18. Raffel, C., Shazeer, N., Roberts, A., Lee, K., Narang, S., Matena, M., Zhou, Y., Li, W., Liu, P.J.: Exploring the limits of transfer learning with a unified text-to-text transformer. J. Mach. Learn. Res. **21**(140), 1–67 (2020)
19. Reimers, N., Gurevych, I.: Sentence-BERT: sentence embeddings using Siamese Bert-networks. In: Proceedings of the 2019 Conference on Empirical Methods in Natural Language Processing (2019)
20. Sagawa, H., Mitamura, T., Nyberg, E.: Correction grammars for error handling in a speech dialog system. In: Proceedings of HLT-NAACL 2004: Short Papers (2004)
21. Shriberg, E.E.: Preliminaries to a Theory of Speech Disfluencies. PhD thesis, University of California (1994)

22. Suhm, B., Myers, B.A., Waibel, A.: Interactive recovery from speech recognition errors in speech user interfaces. In: The 4th International Conference on Spoken Language Processing (ICSLP) (1996)
23. Suhm, B., Myers, B.A., Waibel, A.: Model-based and empirical evaluation of multimodal interactive error correction. In: Proceeding of the CHI'99 Conference on Human Factors in Computing Systems: The CHI Is the Limit (1999)
24. Suhm, B., Myers, B.A., Waibel, A.: Multimodal error correction for speech user interfaces. ACM Trans. Comput.-Hum. Interact. **8**(1), 60–98 (2001)
25. Suhm, B., Waibel, A.: Exploiting repair context in interactive error recovery. In: Fifth European Conference on Speech Communication and Technology (EUROSPEECH) (1997)
26. Vaswani, A., Shazeer, N., Parmar, N., Uszkoreit, J., Jones, L., Gomez, A.N., Kaiser, Ł., Polosukhin, I.: Attention is all you need. In: Advances in Neural Information Processing Systems, vol. 30, pp. 5998–6008. Curran Associates, Inc. (2017)
27. Vinyals, O., Fortunato, M., Jaitly, N.: Pointer networks. In: Advances in Neural Information Processing Systems, vol. 28, pp. 2692–2700. Curran Associates, Inc. (2015)
28. Wang, S., Che, W., Liu, T.: A neural attention model for disfluency detection. In: Proceedings of the 26th International Conference on Computational Linguistic (COLING) (2016)
29. Weng, Y., Miryala, S.S., Khatri, C., Wang, R., Zheng, H., Molino, P., Namazifar, M., Papangelis, A., Williams, H., Bell, F., Tür, G.: Joint contextual modeling for ASR correction and language understanding. In: Proceedings of the IEEE International Conference on Acoustics, Speech, and Signal Processing (ICASSP) (2020)
30. Wolf, T., Debut, L., Sanh, V., Chaumond, J., Delangue, C., Moi, A., Cistac, P., Rault, T., Louf, R., Funtowicz, M., Davison, J., Shleifer, S., von Platen, P., Ma, C., Jernite, Y., Plu, J., Xu, C., Scao, T.L., Gugger, S., Drame, M., Lhoest, Q., Rush, A.M.: Transformers: state-of-the-art natural language processing. In: Proceedings of the 2020 Conference on Empirical Methods in Natural Language Processing (EMNLP): System Demonstrations (2020)
31. Xie, Z., Avati, A., Arivazhagan, N., Jurafsky, D., Ng, A.Y.: Neural language correction with character-based attention. CoRR, abs/1603.09727 (2016)

Learning Affective Responses to Music from Social Media Discourse

Aidan Beery and Patrick J. Donnelly

1 Introduction

The ability of music to elicit powerful emotional responses in listeners is widely recognized across human cultures and societies. Although the connection between human emotions and music is not fully understood, the fact that listening to music engages the limbic system [32] hints at the importance of music throughout our biological and cultural evolutions [16]. In recent years, there has been growing interest in the music information retrieval (MIR) community to design computational approaches capable of estimating human emotional responses to music. If researchers studying music emotion recognition (MER) tasks could automatically estimate typical human responses to any piece of music, music recommendation systems would be able to make more emotionally informed suggestions.

Methods for music mood evaluation typically rely on human annotators to each listen to a musical excerpt and provide a subjective rating based on their perceived emotive response. These subject studies are both expensive and time-consuming, requiring many annotators to rate each song to ensure statistically significant sampling. This dearth of annotated data has hindered the advancement of music emotion recognition systems. Furthermore, no standard for emotion modeling has emerged among researchers. Models of affective response often range from mood classification tasks [10] to prediction of continuous valence-arousal values [14].

Despite these challenges, there have been a variety of attempts to predict a song's emotive qualities automatically. Early methods attempted to learn associations with emotion from manually engineered acoustic features [42]. However, such approaches have been insufficient, and researchers have declared a "semantic

A. Beery · P. J. Donnelly (✉)
Oregon State University, Corvallis, OR, USA
e-mail: beerya@oregonstate.edu; donnellp@oregonstate.edu

M. Abbas (ed.), *Practical Solutions for Diverse Real-World NLP Applications*,
Signals and Communication Technology,
https://doi.org/10.1007/978-3-031-44260-5_6

gap" between low-level acoustic descriptors and perceptual features observed by human listeners [48]. Furthermore, copyright concerns restrict MIR researchers from distributing audio recordings alongside music emotion datasets, hindering the exploration of audio information as a feature space. Lyrics have been used to augment acoustic feature models [34, 63] and by themselves [26, 2]. However, not all music contains lyrics, limiting the generalizability of this approach. Researchers have also explored other modalities, including heart rate [28], electrodermal activity [67], and video of facial expressions [33], often with little success. A few studies have reported limited success with estimating music emotion values from the tags provided by users on online music metadata aggregators such as Last.fm[1] [17, 5, 6]. However, this feature space is relatively small, often only consisting of a few dozen single-word descriptors which members of the Last.fm community deemed relevant to a piece of music.

We hypothesize that the conversations users have about a piece of music might contain semantic clues about typical affective responses to that piece of music. We present a novel approach [20] for learning the continuous valence and arousal values of a song using only the social media conversations referencing that song. To achieve this, we compile a large dataset of social media music discourse using the songlists from four music emotion datasets: AMG1608 [14], PMEmo [67], DEAM [3], and Deezer [17]. We train several large language models to predict music emotion values from these social media comments alone without relying on audio signal analysis or lyrics information. We believe this to be the first approach to estimate the affective qualities of a song solely from social media conversations.

2 Related Work

Music emotion recognition is the task of training computational models to estimate a culturally average emotive response for a piece of music. The ability to automatically understand the relationship between the audio signal of a piece of music and the anticipated human emotion is of great interest to the field of music information retrieval. This is a particularly difficult problem because affective responses about music are subjective and vary both within and between different culturally entrained groups of listeners. Researchers studying this problem typically train models to estimate an emotional response based on the average of multiple different annotators.

Furthermore, there is a large semantic gap between high level music concepts and low-level acoustic features extracted directly from an audio signal. To overcome these difficulties, researchers have explored many different modalities, including descriptive features from audio, music scores, song lyrics, music videos, and even physiological signals monitoring the listener.

[1] https://www.last.fm/

Research studies in music emotion recognition typically seek to classify a categorical label for the entire song with classification [36, 27, 34, 65, 40] or to estimate dimensional values with regression [65, 13, 53, 44, 54]. Researchers have also explored probabilistic mappings between categorical and dimensional emotion semantics of music [58]. These predictions are most typically made at the song-level [27, 43, 39, 15, 19, 11] although there is also active research attempting to track dynamic changes in emotion over time [61, 53, 37, 11].

2.1 Acoustic Features

Traditionally, researchers exploring models to recognize music emotion have relied upon low-level features extracted from the audio signal of a song. Many studies rely upon features extracted from common audio toolkits and frameworks, such as PsySound [9], MARSYAS [56], jAudio [46], YAFFe [45], OpenSmile [22], or Essentia [7]. In other cases, the authors craft customized signal processing methods to attempt to capture information from the audio signal that might be useful in attempts to predict human emotional responses to music. Over the years, researchers have explored thousands of features measuring pitch, melody, harmony, rhythm, dynamics, timbre, and expression (see [48] for a review). Using these descriptive audio features to train machine learning models, researchers have explored many different algorithms, such as linear regression [13, 40, 53], support vector machines [36, 34, 65, 27, 23], support vector regressors [65, 53, 61], random forests [34], and Gaussian models [42, 53], and in recent years, deep learning approaches such as autoencoders [12], generative adversarial networks [29], convolutional neural networks [19, 15], and recurrent neural networks [61, 37, 43, 39, 47, 11].

In one early approach, researchers applied support vector machines in an attempt to classify 13 different emotions, using features extracted from 30-second excepts of 499 audio files across 128 different albums covering four genres of music [36]. The authors reported an F_1 score of only 0.41, highlighting the difficulty of the problem. One major limitation of this study is that the emotion labels were all labeled by a single expert listener. In another study, which considered only a single genre, the authors employed several domain experts to manually annotate a dataset of 250 pieces of classical music with one of four emotions: *contentment*, *depression*, *exuberance*, and *anxiety*. After extracting numerous rhythm and timbre features, the authors applied a Gaussian mixture model to achieve 86.3% accuracy [42]. Acknowledging that annotator fatigue may lead to inconsistent emotion labels, one study designed a music emotion prediction tool to help reduce annotator fatigue in the hopes of yielding more robust datasets [63]. Because emotional reactions to music are highly subjective, another study sought to increase the number of annotations available for each of the 200 songs in their dataset. Crowdsourcing the task online, the authors collected an average of 28.2 annotations across their dataset of 30-second excerpts of film score soundtracks [62], labeling eight different moods: *sublime*, *sad*, *touching*, *easy*, *light*, *happy*, *exciting*, and *grand*. For this task, the study reported a cosine similarity of 0.73 after training support vector machines with acoustic features. Although the authors extracted a total of 88 features, they reported that they achieved similar efficacy with only the best 29 features.

More recently, Chowdhury et al. investigated the development of mid-level features with the hope of helping to close the semantic gap between low-level audio features and human emotive responses to music [15]. These mid-level features describe perceptual concepts, such as tonal stability, articulation, and rhythm. The authors performed feature-importance analysis and trained convolutional neural networks to predict emotions from a dataset of 110 movie soundtracks, achieving a correlation of 0.71 relative to annotations by experts. In general, intelligent systems have struggled to predict human responses to music based on acoustic features alone. There remain disconnections between audio descriptors and high level music concepts. Because of this semantic gap between low-level audio features and human affective responses, researchers are limited in their ability to predict emotional response from acoustic information alone [48]. To improve the prediction of affective responses from audio, it seems necessary to supplement audio features with additional modalities [64].

2.2 *Natural Language Processing Approaches*

Given the predictive limitations of learning from audio alone, researchers considered the potential of song lyrics to aid in the prediction of the emotional qualities of a song. Investigators first began by examining statistical correlations between features extracted from the audio and the lyrics as well as the relationship between these features and the emotion annotations themselves [44]. To compensate for the lack of annotated data, one study synthetically generated emotion labels for a dataset of 100 pop-genre songs. They extracted popular tags from Last.FM and compared against the lexical database WordNet.[2] They applied latent semantic analysis, training self-organizing maps to annotate songs with four mood categories (*angry*, *happy*, *sad*, *relaxed*), manually verifying over two-thirds of labels [35]. The authors reported lower accuracy using lyrics alone (62.5%) compared to the models built on acoustic features (89.8%) [34]. The authors found that by combining acoustic and lyric features together, they were able to increase accuracy by three percent (92.4%) [5].

In a sequence of studies, Hu and Downie examined the relationship between emotion labels and text-based features extracted from the lyrics. To annotate their dataset with synthetic annotations of 18 moods, the authors used Last.FM tags and their WordNet distance to words in the ANEW word list [8] in order to estimate valence, arousal, and dominance values for 5,296 songs in their dataset [27]. The authors then compared various approaches to lyric sentiment analysis [25] in order to identify cases in which the performance of lyric-only models exceeded those

[2] https://wordnet.princeton.edu/

of acoustic feature models [26]. Overall, the authors found their lyric-only model (63.7%) outperformed their audio-based model (57.9%). A fusion model combining text and audio features showed moderate improvement (67.5%) over lyrics alone. Similarly, another study reported that a late fusion of audio features and text-based features derived from the lyrics improved accuracy of their models from 46.6% to 57.1% [65]. More recently, researchers have investigated the performance of emotion recognition models based solely on the lyrics. In one such study, the authors estimated the valence and arousal values of the words in the lyrics using established word lists to create a song-level prediction of valence and arousal. The authors reported a 74.3% classification accuracy relative to the All Music Guide[3] mood tags [10].

2.3 *Deep Learning Approaches*

Following the many advances in deep neural algorithms and architectures over the last decade [52], researchers have begun exploring music emotion recognition tasks using both acoustic features and text-based lyrics using deep learning.

2.3.1 Deep Learning on Acoustic Features

Researchers have investigated different deep neural architectures to attempt music emotion prediction using acoustic features. For this task, recurrent neural networks outperform feedforward neural networks [61]. Among recurrent architectures, bidirectional long short-term memory (BLSTM) models appear to improve prediction of musical affect over unidirectional long short-term memory (LSTM) models [37]. Additionally, researchers have reported that attentive-LSTM models improve prediction performance of arousal and valence estimations over baseline LSTM models without attention [43, 11]. Bypassing preprocessing and feature extraction altogether, one research team trained bidirectional gated recurrent units directly on raw audio to attempt to classify discrete music emotions [47].

One study designed a custom experimental pipeline that makes use of both convolutional and recurrent neural networks. The authors employed a convolutional neural network to learn to select which acoustic features were subsequently used to train an LSTM model. On a custom dataset of 30 second excerpts of 124 pieces of Turkish traditional music, the authors achieved classification accuracy of 92.7% for three broad categories of emotion, which outperformed their baseline algorithms of support vector machines, random forest, and k-nearest neighbor [24]. Another recent study proposed adapting a generative adversarial network with double-channel attention mechanism (DCGAN) in order to learn the dependence

[3] https://www.allmusic.com/

between music features across channels [29]. To evaluate their architecture, the authors designed an experiment classifying five emotional characteristics (*happy*, *sad*, *quiet*, *lonely*, *longing*) on a custom dataset of 637 songs, reporting that the DCGAN (89.4%) outperformed both convolutional and recurrent architectures.

2.3.2 Deep Learning on Lyrics

In addition to the approaches to estimate emotional responses to music directly from the audio of a song, other researchers have studied the use of the lyrics of a song using deep learning to estimate human responses to music. Using only the text of the song lyrics, Agrawal et al. trained the `xl-net` transformer [66] and achieved around 95% classification accuracy on a large dataset of lyrics [2]. This encouraging result may imply large language models have the ability to capture meaningful semantic relationships from music lyrics without additional acoustic descriptors.

More recently, investigators have begun adapting large language models (LLM) and algorithms to learn embeddings directly from the audio signal. One recent study combined a large-scale pretrained language model with an audio encoder to attempt to generate interpretations from cross-modal inputs of song lyrics and musical audio [68]. Another research team explored representations of music using a joint embedding of natural language and audio [30]. Using these embeddings of over 5,000 music and text pairs, the team trained generative acoustic models able to produce music based on a text description given as input [1].

2.4 Large Language Models

Transformers are deep learning models based on the principle of self-attention [57]. This LLM architecture, first introduced in 2017, has quickly become popular in the areas of natural language processing (NLP) and computer vision (see review [38]) where large pretrained models have achieved state-of-the-art performances in a wide variety of tasks. In this section, we briefly review the four transformer-derived models that we compare in this study.

2.4.1 BERT

BERT, or Bidirectional Encoder Representations from Transformers, is a popular transformer model for learning representations of natural language [18]. BERT leverages a large dataset of unstructured English text from Wikipedia and assorted literature. By taking unlabeled sequences of English text, corrupting parts of the input, and attempting to predict the missing tokens, the model encodes complex relationships between words and demonstrates the ability to learn robust language representations. This self-supervised pretraining objective function is referred to as masked language modeling and has become the foundation for many similar models. Because BERT and other LLMs are pretrained on very large amounts of data, these models can be fine-tuned to new tasks relatively quickly. However, model training still requires significant compute resources, especially when learning large datasets. BERT is widely used in many NLP tasks, including machine translation, dialogic generation, question answering, and sentiment analysis.

2.4.2 DistilBERT

DistilBERT seeks to address the immense computational requirements of BERT while retaining the same capability to learn effective language representations. To accomplish this, they leverage knowledge distillation to train a smaller model to emulate the behavior of BERT [51]. By optimizing DistilBERT to predict the same output probabilities as BERT during pretraining, the authors design a model which retains 97% of the performance on benchmark NLP tasks. DistilBERT bases its architecture on BERT but reduces the number of hidden layers from 12 to 6. This lowers the number of model parameters by 40%, enabling faster training and fine-tuning.

2.4.3 RoBERTa

RoBERTa aims to surpass the performance of BERT by both leveraging a larger pretraining corpus and by modifying the pretraining objective task [41]. The authors replicate the architecture of BERT while empirically studying how various factors in the pretraining of large language models impact downstream task performance. They find that encoder transformer models respond positively to significantly larger pretraining datasets and longer pretraining schedules with larger batch sizes. Furthermore, they propose a dynamic masking method, randomly altering which tokens are masked during the masked language modeling task, which helps to improve the performance of RoBERTa on NLP benchmarks compared to BERT.

2.4.4 `xl-net`

`xl-net` is another transformer architecture which seeks to improve upon limitations of large transformer models with autoencoder-based pretraining objectives such as BERT and RoBERTa [66]. The `xl-net` architecture features an autoregressive objective function, doing away with the masked language modeling approach used by BERT and its derivatives. This, the authors propose, should reduce the asymmetry between the distributions learned during the pretraining process and those of the inputs in downstream tasks. Unlike previous autoregressive approaches, `xl-net` models all possible permutations of the input sequence, enabling it to learn long-range and bidirectional dependencies of words in context. Like other LLMs, it is pretrained on a large corpus of text and subsequently fine-tuned with additional training for specific NLP tasks. `xl-net` has been shown to achieve competitive results on many NLP benchmarks.

3 Music Emotion Datasets

In this work, we study the task of music emotion recognition over four datasets of songs with song-level annotations of valence and arousal. These datasets were created using similar procedures: tasking human annotators to listen to a musical excerpt and provide a numeric description of their emotive response. The labor-intensive nature of collecting these annotations has hindered research in music emotion prediction. The advent of crowdsourcing platforms has enabled experimenters to reach a wider audience; however, these annotations are still expensive and time-consuming [14]. Researchers have also created synthetic valence-arousal annotations [17] by mapping community-provided features, such as Last.fm tags or metadata from the All Music Guide, to existing word-affect datasets [60].

3.1 AMG1608

The AMG1608 dataset provides 1,608 songs selected from the All Music Guide (AMG) and rated for valence and arousal by 665 annotators [14]. The dataset's creators aimed to develop a large and state-of-the-art music emotion recognition dataset. To conduct an annotation experiment at this scale, the authors used Amazon Mechanical Turk—an online crowdsourced work platform—to reach a large subject pool. AMG users rated songs for 34 different mood categories, which are converted from mood labels to valence-arousal estimates using the `tag2VA` algorithm [59]. From this, a subset evenly distributed in the valence-arousal space was selected. 665 annotators participated in the experiment, and between 15 and 32 annotations were collected per song. 46 annotators provided ratings for over 150 songs, presenting a unique opportunity for a study of emotion-aware music recommender personalization as well as introducing a potential bias in the label distribution. Each annotator listened to a 30-second excerpt of the sample and provided a dimensional rating in the circumplex model of emotion. Each coordinate was treated as a valence-arousal label, and individual coordinate labels were averaged between annotators to produce an emotion label for a given song.

3.2 *PMEmo*

Music emotion recognition datasets typically fall into one of two categories: small datasets with few annotators rating samples in lab environments to yield high-quality individual annotations or larger datasets with many annotations of relatively lower quality gathered using online crowdsourcing platforms such as Amazon Mechanical Turk. The PMEmo dataset fills the need for music emotion datasets with both high-quality annotations and many samples by conducting a large-scale human subject study in a laboratory setting [67]. 457 annotators, including 366 Chinese university students, participated to annotate valence and arousal over a collection of 794 songs.

1,000 songs were selected from record label industry charts between 2016 and 2017, such as Billboard Top 100, iTunes Top 100, and UK Top 40 Singles. After deduplication, 794 songs remained, primarily representing Western pop music. A 30-second sample representing the chorus of each song was manually excerpted by music students. Annotators were instructed to listen to each sample and provide dynamic valence-arousal ratings at a 2 Hz sample rate using an annotation interface. At the end of the sample, annotators were then asked to provide a single valence-arousal rating representing their overall emotive response to the song. These static annotations were averaged to provide valence-arousal labels for each song. Electrodermal activity was also recorded from participants during the listening and annotation experiment (Fig. 1).

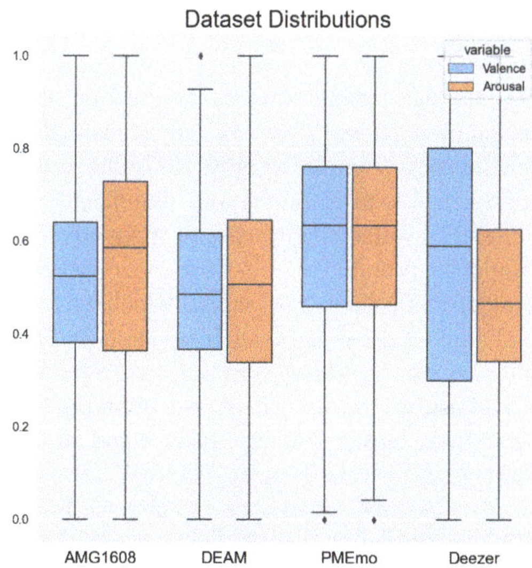

Fig. 1 Distributions of the valence-arousal labels for each of the datasets

3.3 DEAM

Research in music information retrieval continues to be hindered by a lack of annotated datasets with accompanying audio data. Copyright restrictions on the majority of publicly released music prohibits researchers from distributing audio recordings in music datasets, posing significant challenges to approaches that learn from audio data. In response, Soleymani et al. provide a dataset of royalty-free songs annotated for affective qualities using continuous emotion labels [3].

The DEAM dataset consists of 1,803 songs from `freemusicarchive`, `jamendo`, and medleyDB, online repositories of royalty-free music. 45-second segments were selected randomly from each sample, and annotators were asked to provide dynamic valence and arousal annotations at a 2 Hz sample rate while listening to this excerpt. These dynamic ratings were averaged over time to provide a single per-participant valence-arousal annotation. Annotators were recruited from Amazon Mechanical Turk, and each song received a minimum of five annotations. Along with averaged valence-arousal annotations, the authors provided both the excerpt and full-length audio for each song. Although the royalty-free licensing of these songs permits the distribution of the audio recordings, it seems that these copyright-free samples are more obscure than songs included in other datasets. For this reason, we expect to find less online discourse about these songs compared to popular songs used in other datasets.

3.4 Deezer

Despite these efforts towards the creation of large-scale annotated music emotion recognition datasets, even the largest manually annotated datasets only consist of a few thousand samples at most. To evaluate the utility of deep learning approaches for valence-arousal estimation, significantly more data is necessary than what is currently available. The cost of manually annotating datasets at sufficient scale would be prohibitive. Researchers at Deezer developed a dataset consisting of synthetic valence-arousal labels, which we refer to as the Deezer dataset [17].

In this study, songs available in both the Million Song Dataset [4] and Deezer's music streaming library were selected. Each song's associated tags were aggregated from Last.fm, providing a list of community-provided key descriptors for each song. From these tags, a synthetic valence and arousal label was generated by comparing the tags against the Extended ANEW dataset [8], a collection of 14,000 English words annotated for valence and arousal [60]. The Deezer dataset operates on the fundamental assumption that the valence and arousal annotations of English words from Warriner et al.'s experiments transfer to the music emotion space, and that the descriptors added by community users on Last.fm meaningfully relate to the song's emotive qualities. For these reasons, the authors concede that their dataset is not as robust of a ground truth as manually annotated music emotion datasets (Fig. 2).

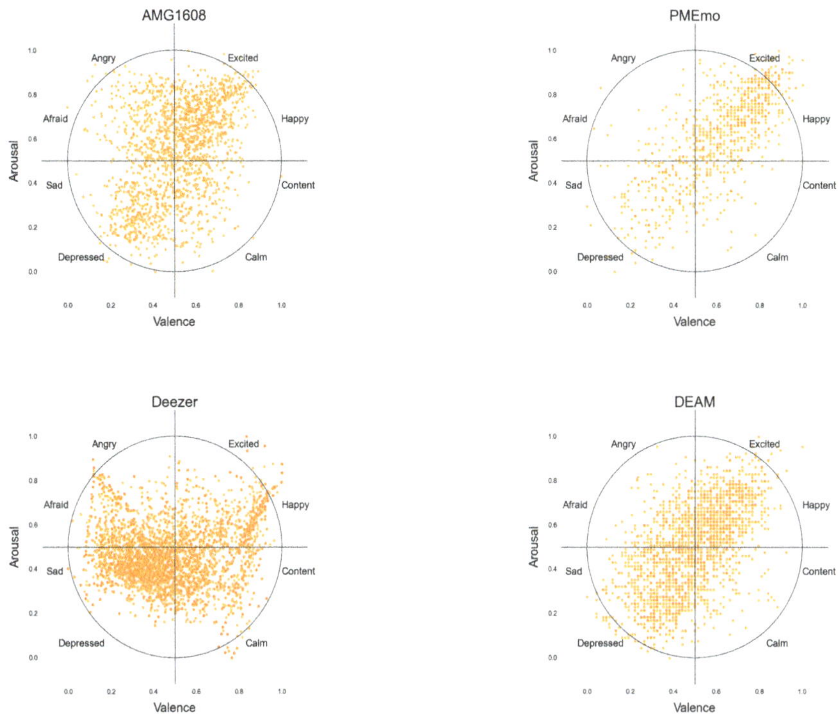

Fig. 2 Circumplex models, representing distributions of the labels from four music emotion datasets across Russell's emotional space [50]

4 Musical Discourse Model

We propose a system for the automatic prediction of static valence and arousal targets using the social media discourse related to a song to estimate users' average emotive response to that song. To accomplish this, we collect social media commentary from Reddit[4] and YouTube,[5] both platforms with active music subcultures engaging in discussion about music. From these conversations alone, and without considering the song's audio or lyrics, we attempt to predict a song's valence and arousal by fine-tuning pretrained large language models. We focus our investigation on transformer-based encoder models, such as BERT [18] and its derivatives, on a two-target regression task of estimating song-level estimates of music emotion from online comments associated with a musical sample.

4 https://www.reddit.com/

5 https://www.youtube.com/

Table 1 Details of the valence and arousal labels in selected MER datasets

Dataset	Songs	Label type	Scaling
AMG1608	1,608	Crowdsourced	[−1, 1]
DEAM	1,803	Crowdsourced	[0, 10]
PmEmo	767	Lab Survey	[0, 1]
Deezer	18,648	Synthetic	[−3, 3]

4.1 Collecting Social Media Commentary

We collect social media discourse which references the songs from AMG1608 [14], PMEmo [67], DEAM [3], and Deezer [17]. In total, our dataset gathers social media comments about 19,627 songs, of which 4,179 are manually annotated. For each sample in the four datasets, we query the two social media platforms for posts which make direct reference to both the song title and the artist. From each platform, we select the 50 highest rated submissions as ranked by each platform's search API. We then collect every comment and reply which corresponds to any of these top-level submissions. If a song has not been discussed on Reddit or YouTube, we omit that song from our discourse dataset. Table 1 describes our dataset of retrieved comments.

We collected data over a six month period between November 2022 and April 2023, scraping both past and recent comments. We achieved higher retrieval rates from YouTube across all four songlists; 84% to 97% of all queried songs had at least one matching post. Retrieval rates from Reddit were lower, although we found reference to over 80% of songs from AMG1608 and PMEmo. However, only 11% of songs from DEAM had corresponding comments on Reddit. When comparing DEAM against similarly sized datasets, we observed that our data collection totals 1,472,021 and 881,931 comments for the songs in AMG1608 and PMEmo datasets, respectively, but only 303,667 comments reference songs from DEAM. In total, our dataset of musical discourse contains more than 11 million comments. Figure 3 shows the distribution of the retrieved comments and their associated lengths. Unsurprisingly, the popular songs from the PMEmo dataset were associated with higher rates of discourse. Conversely, for the relatively obscure songs in the DEAM dataset, we found significantly fewer comments per song than other datasets (Table 2).

4.2 Model Design

We evaluate our musical discourse dataset as a feature space in the task of music emotion recognition by applying large pretrained transformer models designed for natural language understanding tasks to our corpora. These models learn language representations from very large datasets of unstructured text using self-supervised language modeling tasks. From this pretraining, these models can be fine-tuned to

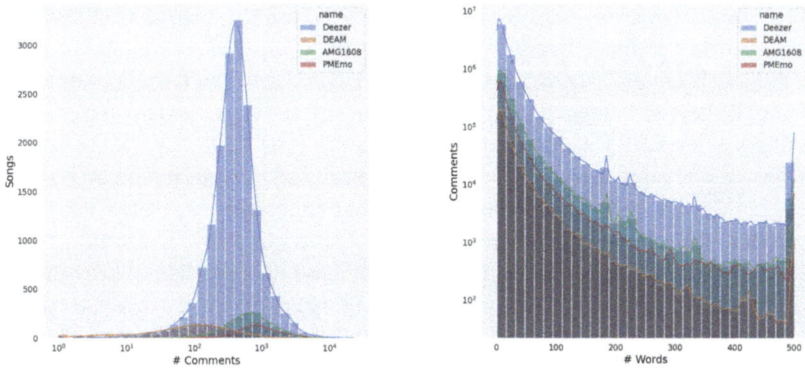

(a) (b)

Fig. 3 Social media discourse dataset distributions, stratified by dataset. (**a**) Number of comments per song. (**b**) Length of comments

Table 2 Summary statistics for our dataset of social media commentary by source

		Songs		Comments			Words	
		n	Yield	n	μ	σ	μ	σ
AMG1608	Reddit	1,412	88%	578,283	409.5	1,180.5	15,796	43,847
	YouTube	1,563	97%	893,738	571.8	268.1	11,424	5,917
PmEmo	Reddit	624	81%	391,325	627.1	1,122.1	21,065	47,179
	YouTube	736	96%	490,606	666.6	267.1	11,333	5,806
DEAM	Reddit	205	11%	69,943	341.2	1,873.1	13,562	67,127
	YouTube	1,508	84%	233,724	155.0	194.8	3,153	5,030
Deezer	Reddit	11,122	60%	2,497,517	224.6	767.2	8,963	30,649
	YouTube	16,435	88%	6,685,202	406.8	229.2	7,524	4,792
Total		19,627	86%	11,840,338	603.3	911.5	14,904	32,125

learn downstream tasks in relatively few epochs [18, 41]. We fine-tune one such model, BERT, on our multi-target regression task by assigning each song's valence and arousal label to the comments relating to that song, and attempting to predict a song's emotive annotations directly from this social media discourse. We then compare the performance of BERT on this task to a selection of other pretrained large language models: DistilBERT [51], RoBERTa [41], and `xl-net` [66].

We use the implementations of these large language models provided by the Hugging Face deep natural language processing library.[6] Each input consists of a single social media comment, labeled by the valence and arousal annotation for the song which the comment is associated with. We tokenize each input using the `TokenizerFast` library and use a maximum sequence length of 128 tokens, truncating and right-padding inputs to coalesce sequences to this dimension.

[6] https://huggingface.co/models

Our model consists of two components: a pretrained transformer to learn a representation for input comments, and a fully connected neural network with one hidden layer to learn a regression target from this language representation. We output the last hidden state of the `[CLS]` token, as is standard for designing classifiers using BERT's language representations [18]. Our fully connected layer, serving as the regression head, learns a valence and arousal output based on these `[CLS]` last-hidden-state vectors. We fine-tune BERT to adapt the learned representations to our downstream task. This does incur a risk of overfit, as BERT and its derivatives have significantly more parameters than our dataset has examples. We limit our fine-tuning to two epochs to mitigate overfit as recommended in [18, 51]. We use a mean-squared-error loss and the Adam optimization algorithm [31] with a learning rate of 1×10^{-5}.

4.3 Experimental Design

We randomly partition each songlist into training, validation, and test subsets with $0.70 \times 0.15 \times 0.15$ split, respectively. All comments associated with a song are then placed in that song's corresponding subset. Valence and arousal labels are normalized and scaled to [0, 1]. Inputs are filtered to remove URLs and HTML tags. Further text preprocessing is unnecessary for fine-tuning of pretrained large language models, as BERT and similar models use unfiltered text from online sources for their pretraining tasks [18]. These models expect inputs to adhere to standard grammatical structure, and as such we do not lemmatize nor remove stopwords from our comments. Each comment is assigned a music valence and arousal prediction from our regression model. To produce song-level valence and arousal labels from these comment-level outputs, we take the average of all output labels for each comment associated with a song to produce the final valence-arousal estimation for that song.

5 Music Emotion Recognition with Large Language Models

We test the performance of a BERT-based regression model for predicting the emotive qualities of a piece of music. We begin by measuring the performance impact of language models trained to be case-sensitive versus their uncased counterparts. Next, we test comment-level filtering schemes, evaluating the impact of dropping short comments or those below a certain score threshold, where score is a measure of likes or upvotes from the comment's platform of origin. We compare several large language models to investigate the impact of different transformer architectures and pretraining schemas for this task. For our tuning experiments, we test each model configuration on both AMG1608 and PMEmo, chosen for their manually annotated labels and active discussion on the social media platforms we investigate. We select the filtering strategy and pretrained model with the best performance on these two datasets and investigate its performance on the DEAM and Deezer datasets.

5.1 Model Parameters and Dataset Preprocessing

We compare model implementation and dataset preprocessing methods for predicting music emotion targets from AMG1608 and PMEmo using our musical discourse dataset. First, we evaluate both cased and uncased versions of BERT for this task and measure the performance implications of case sensitivity in BERT pretraining on our task. Informed by this experiment, we then test dataset filtering methods to identify a preprocessing strategy to reduce potential noise in our social media data.

5.1.1 Case Sensitivity of Language Model

We explore two versions of the BERT model: one case-sensitive (`bert-base-cased`) and the other, case-insensitive (`bert-base-uncased`). `bert-base-uncased` uses the same pretraining tasks as its cased counterpart. However, during pretraining, all text is transformed to lower-case and accent markers are removed.[7] We compare the performance of these two models on our task. For each model, we run experiments on AMG1608 and PMEmo. In both cases, the model is trained on a combination of comments from YouTube and Reddit.

In Table 3 we show the Pearson's correlation between our models' predictions and the datasets' annotated labels. When training on AMG1608, we observe a slight yet measurable improvement in model performance with the cased variant of BERT. We expect this improvement results from the use of capitalization as an important mechanism for conveying tone or intent and, therefore, provides important semantic information for our emotion recognition tasks. We use `bert-base-cased` in following experiments.

Table 3 Pearson's correlation of cased and uncased variants of BERT

	PMEmo		AMG1608	
	Valence	Arousal	Valence	Arousal
bert-cased	0.68	0.47	0.51	0.75
bert-uncased	0.68	0.49	0.45	0.74

[7] See https://huggingface.co/bert-base-uncased

5.1.2 Comment Filtering

Though we perform some preprocessing on the input text, this does not mean that all inputs are useful for our downstream task. To address the innate noisiness of social media data, we evaluate a selection of strategies for rejecting certain comments from our training set. We begin by filtering comments based on the number of likes or upvotes they receive from other users on their respective social media platform. We assume that highly rated comments are more likely to express sentiments shared by the community. By filtering out comments with lower scores, we hope to prune off-topic discussion and spam, as these types of responses are less likely to be informative. We initially begin with a score threshold of 3, based on the criteria used by other large language models which rely on Reddit data for their pretraining dataset [49]. This aggressive filtering method removes a total of 71% of all comments across AMG1608 and PMEmo. We also explore a weaker filtering threshold, requiring only that the score be positive ($\geq$1), excluding a more modest 36% of our total comments.

When training on data filtered with a score threshold of 3, we observe marginal improvements in performance over the unfiltered BERT baseline across all dimensions except valence on PMEmo, for which we observe a decrease in correlation by 28%. The less aggressive score threshold of 1 does not appear to have as drastic impact on performance on the PMEmo dataset and in fact outperforms baseline by 4% to valence and 10% to arousal. However, on the AMG1608 dataset, lowering the score threshold weakens performance overall and it does not exceed baseline performance. We suspect that dataset size is an important factor in this large difference in per-dataset performance. A filtering regimen which works well for AMG1608 may remove too many comments from PMEmo, and one which optimizes for PMEmo may leave too many noisy inputs for AMG1608. Additionally, we test filtering comments by length. Our model expects inputs of at most 128 words, adding zero-tokens to pad all inputs to this dimension. Though attention masking allows our model to handle zero-padded inputs without introducing excess noise, longer inputs contain more semantically meaningful tokens in each input tensor. We identify that the bottom quartile of comments in the combined AMG1608 and PMEmo musical discourse dataset has at most 30 characters. We drop this lower quartile of comments.

We combine our score threshold and length threshold filters to require all comments both be longer than 30 characters and have a positive score. Comparatively, this yields improved performance when predicting arousal labels for PMEmo and valence labels for AMG1608 than any other individual preprocessing method. However, this comes at the cost of a reduction in predictive performance in the other dimensions. When we filter short comments in the bottom quartile of character length, we achieve the best performance overall, with a 2.1% increase over baseline. We apply this technique in our comparison of language models in Sect. 5.3 (Table 4).

Table 4 Impact of comment filtering strategies on model performance

	PMEmo			AMG1608		
	Valence	Arousal	Count	Valence	Arousal	Count
Baseline	0.68	0.46	688,712	0.51	0.75	1,281,473
Score ≥ 3	0.49	0.47	221,360	0.53	**0.79**	352,481
Score ≥ 1	0.71	0.51	448,631	0.45	0.74	822,623
Length ≥ 30	**0.80**	0.48	474,827	0.52	0.65	976,081
Joint filtering	0.63	**0.57**	334,215	**0.62**	0.57	654,316

Table 5 Performance of source-specific models

	PMEmo		AMG1608	
	Valence	Arousal	Valence	Arousal
YouTube	0.54	0.47	0.60	0.65
Reddit	0.51	0.50	0.34	0.54
All	0.80	0.48	0.52	0.65

5.2 Comparison of Social Media Sources

Next, we compare the utility of conversations from each social media platform for the purpose of music emotion recognition in Table 5. YouTube is widely used for sharing and listening to music, and we anticipate that users commenting most likely recently watched the music video. Reddit, on the other hand, is more suitable for longer conversations but lacks the YouTube's popularity as source for music. We believe both types of conversations contain semantic information relevant to the task of music emotion recognition. However, the difference in both user experience and intent warrants an investigation into models trained on individual sources.

We find that a model trained on exclusively YouTube comment data exceeds our baseline model's performance on the AMG1608 dataset. However, neither source-specific model outperforms the best scores for AMG1608 presented in Table 4. A YouTube specific model underperforms relative to our multi-source baseline on PMEmo dataset, again demonstrating that additional filtering adversely impacts performance for datasets lacking large amounts of commentary. We find that models trained on only YouTube conversations outperform those trained only on Reddit data, which we attribute to the fact that our dataset includes 135% more YouTube comments than Reddit submissions. Overall, our combined source model outperformed both models trained on individual social media sources.

5.3 Comparison of Language Models

As detailed in Sect. 2.4, various BERT-derived models have sought to address limitations with the model and improve its performance on downstream tasks. We compare three of these models, fine-tuning each on our emotion prediction task.

Table 6 Performance of selected pretrained large language models after fine-tuning

	PMEmo		AMG1608	
	Valence	Arousal	Valence	Arousal
BERT	0.80	0.48	0.52	0.65
DistilBERT	0.80	0.46	0.49	0.64
RoBERTa	0.79	0.53	0.55	0.67
`xl-net`	0.80	0.50	0.55	0.63

Table 7 Pearson's correlations to the ground truth labels for the four datasets

	Valence	Arousal
AMG1608	0.55	0.67
PMEmo	0.79	0.53
DEAM	0.08	0.02
Deezer	0.47	0.43

Specifically, we investigate DistilBERT [51], RoBERTa [41], and `xl-net` [66], each addressing different limitations of the original approach (Table 6).

We observe a small improvement in performance using language models pre-trained on larger corpora of text. Predictions generated from models using DistilBERT achieve correlations within 97% of baseline, reinforcing claims made in [51]. `xl-net` and RoBERTa outperform BERT by 1.3% and 4.3%, respectively. Both of these models use significantly larger datasets for their respective pretraining approaches. We select RoBERTa for our final model evaluations in Sect. 5.4.

5.4 Dataset Comparison

In our previous experiments, we focused on the PMEmo and AMG1608 datasets because the songs in these datasets were annotated by human subjects. In our final experiment, we compare our approach's performance across the available datasets, including both DEAM, which suffers from a lack of relevant social media commentary, and Deezer, whose annotations were synthetically generated. In this experiment, we first fine-tune RoBERTa using the combined commentary from both Reddit and YouTube. We then filter all comments shorter than 30 characters, dropping the shortest 25% of comments. We evaluate this model's performance on the DEAM and Deezer datasets, whose distributions of labels differ from those of AMG1608 or PMEmo (see Fig. 1) (Table 7).

We find that our model's predictions achieve weak but measurable correlations to the synthetically annotated labels in the Deezer dataset. This is not surprising since the synthetically generated annotation is less likely to reflect the range and nuance of affective responses reported by human subjects when listening to music. We hypothesize that the social media discourse our model uses to predict a song's emotive properties may not correlate well with the word-level representations of text tags used to synthesize the Deezer dataset. Although we expected our model to struggle with the DEAM dataset, we observe that our model completely fails to predict emotional responses to the songs from DEAM. The labels in the DEAM dataset are provided by crowdsourced human annotators in a method similar to the labels provided in AMG1608. However, as shown in Table 2, our data collection of social media commentary yields significantly fewer comments associated with songs from DEAM than any other datasets. The average number of YouTube comments per song in DEAM is only 155, compared to between 400 and 650 comments per song for the other datasets. Given the insufficient quantity of discourse, our model was unable to learn from the DEAM dataset.

6 Discussion

We present a novel approach to predict dimensional music emotion labels through sentiment analysis of social media conversations discussing a piece of music. To assess the potential for a model to estimate a song's affective qualities solely from social media discourse, we create a large corpus of online conversations related to the songs in four published datasets for music emotion recognition. We construct a music emotion prediction system using pretrained large language models, leveraging the language representations learned by these transformer models to fine-tune each to this task. Overall, we observe modest correlations between the predictions made by these models and the dimensional emotion labels provided by human annotators.

6.1 Limitations

We visualize our output model predictions in the valence-arousal space in Fig. 4. Our model's predictions tend to cluster closely to the center of this space. This indicates that, despite a moderate linear relationship between our estimates and true labels, these predictions are often collapsing to the average within the distribution. This issue persists despite our attempts to reduce the noise in our dataset. We anticipate this phenomenon results from our approach to aggregating comment-level predictions to generate a single overall valence and arousal estimate for a song. In our approach, we predict a valence and arousal value for each comment, then average the comment-level predictions to produce a song-level estimate. This process discards valuable semantic information. Comment-level predictions cannot capture the relationship between the comments, including those that serve as replies to another. Furthermore, this aggregation process may collapse comments belonging to the same song with conflicting sentiments to a neutral value.

Our model requires songs to correspond to a sufficient volume of comments available on the social media platforms. This requirement restricts our model's ability to make inferences about the emotive qualities of newly released songs

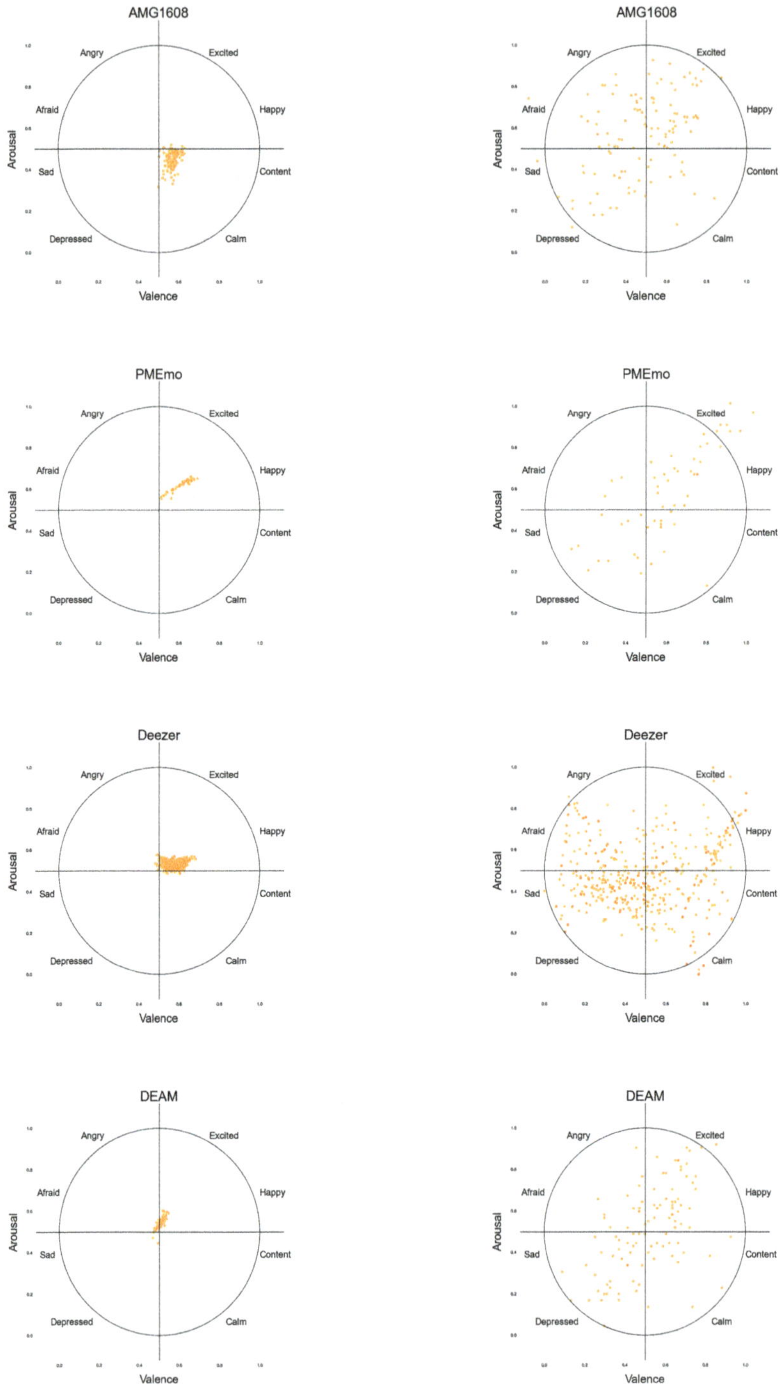

Fig. 4 Comparisons of the distributions of the ground truth labels (right) against our model's predictions (left) in the valence-arousal space

or those that belong to a particularly niche genre. Models which use the audio information and lyrics of a song to make these predictions would not share such limitations. In our analysis of the DEAM dataset [3], we find that the copyright-free nature of these songs was correlated with our inability to find relevant conversational activity online. Without sufficient data, our model was unable to make meaningful predictions.

6.2 Future Work

We will explore methods to preserve the relationships between comments associated with the same song. In the current work, our model expects a single comment, paired with an accompanying music valence-arousal label, for each input. However, this does not reflect the task we seek to learn: a single valence-arousal prediction for a song given a set of comments. In future work, we will investigate new model architectures capable of receiving a single song, with all accompanying discourse packed into a single input tensor, and learning a music affect label without reliance on an aggregation of individual comment-level predictions. A potential approach to address this problem is to simply concatenate comments together to form a single input. BERT and its derivatives use `[SEP]` tokens to indicate the beginnings and ends of distinct sequences within a single input. However, the runtime complexity of transformer models scales quadratically with input size [18], and most large language models limit to a maximum of 1,024 tokens per input. This restricts how many comments can be included as input. Further investigation is needed to design a custom architecture to learn across the different comments of a single thread.

We observe potential in our joint filtering approach in Table 4. Filtering comments in our musical discourse dataset marginally improves model performance. Furthermore, it appears that filtering by both length and score improves performance along dimensions on which our model underperforms, namely PMEmo arousal and AMG1608 valence prediction. However, these filtering strategies resulted in inconsistent performance between datasets, which results from the differences in number of comments available by dataset. By applying dynamic filtering methods, which adjust score and length thresholds based on the number of samples which exist in a dataset, we may address the inconsistent effects of filtering techniques across datasets. Additional criteria for comment filtration should also be introduced and compared against the methods we demonstrate. For example, comments not associated with the expression of an effective response, as determined by existing dictionaries of affective terms [60], could be removed to filter out comments of neutral sentiment.

Additionally, we plan to explore additional sources for music-relevant online discourse. Last.fm and SoundCloud[8] are both online platforms that focus on music,

[8] https://soundcloud.com/

and these communities are a potential source of information directly related to specific pieces of music. The community-provided tags on Last.fm have been used in prior music emotion recognition experiments [35, 17, 6, 14]. This platform also allows users to post comments in response to a specific track with comment mechanism known as “Shouts.” Similarly, SoundCloud users can respond to a song with a public comment. Because these comments stem from communities on music-specific platforms and are in direct response to a song as opposed to responding to a post about a song, we anticipate that the conversations on these platforms may be valuable to a social media-based music emotion recognition system.

To address the cases of songs with limited or no presence on these social media platforms, we intend to explore feature spaces beyond social media data to augment our existing approach. We expect the inclusion of lyrics, song metadata, and acoustic features in conjunction with our social media information to yield a more robust estimator. We hope that the exploitation of these feature spaces will improve a model’s performance on songs for which there is comparatively little online conversation.

7 Conclusion

The development of an automatic system for estimating the emotive qualities of a piece of music has been impeded by a lack of large, high-quality, annotated music emotion recognition datasets. Such annotation experiments are expensive and time-consuming to perform. Furthermore, the distribution of the audio samples used in such datasets is prohibited by copyright law in many cases, restricting the use of an important feature modality for music emotion prediction tasks. We demonstrate the feasibility of predicting continuously valued music emotion labels using only musical discourse from social media platforms. Such an approach does not rely on a song’s audio nor its lyrics, enabling inference to be drawn about a song’s affective qualities indirectly and without access to copyrighted information.

We create a large dataset of social media conversations about musical samples using the songlists provided by four music emotion recognition datasets. In total, we gather over 11 million comments discussing nearly 20 thousand songs. We use this dataset to design and evaluate a system for music emotion prediction using pretrained transformer models. We find that, with relatively few training epochs, these large language models can fine-tune to our music valence-arousal prediction task and provide emotion estimations with moderate correlation to human-provided annotations. To our knowledge, this is the first attempt to predict musical valence and arousal labels using exclusively conversational data from social media platforms.

The ability to predict how an average listener may respond to a piece of music could be used to improve existing music recommender systems. Without the need for costly annotation experiments or licensing for song audio, large music libraries could be rated for dimensional emotional values. The granularity afforded by continuous valence-arousal annotations would allow music streaming services to categorize songs with greater respect to affective characteristics. As another potential broader impact of this research, the ability to quickly and autonomously annotate large libraries of music would enable intelligent systems to automatically generate affect-aware music playlists [21], with potential uses in music therapy [55].

We demonstrate that the conversations people have online about a piece of music can be used to train a model to predict the average affective response elicited in a listener by that song. Our model achieves moderate performance on the prediction of human annotated music emotion targets. Without access to song audio, pre-computed acoustic features, or song lyrics, we are able to fine-tune a large language model to estimate valence and arousal labels corresponding to affective response to a piece of music using only this online musical discourse.

References

1. Agostinelli, A., Denk, T.I., Borsos, Z., Engel, J., Verzetti, M., Caillon, A., Huang, Q., Jansen, A., Roberts, A., Tagliasacchi, M., Sharifi, M., Zeghidour, N., Frank, C.: MusicLM: generating music from text (2023). https://doi.org/10.48550/arXiv.2301.11325. ArXiv:2301.11325 [cs.SD]
2. Agrawal, Y., Shanker, R.G.R., Alluri, V.: Transformer-based approach towards music emotion recognition from lyrics. Adv. Inf. Retr. (ECIR) **12657**, 167–175 (2021). https://doi.org/10.1007/978-3-030-72240-1_12. ArXiv: 2101.02051
3. Aljanaki, A., Yang, Y.H., Soleymani, M.: Developing a benchmark for emotional analysis of music. PloS one **12**(3), 1–22 (2017). https://doi.org/10.1371/journal.pone.0173392
4. Bertin-Mahieux, T., Ellis, D.P.W., Whitman, B., Lamere, P.: The million song dataset. In: Proceedings of the 12th International Conference on Music Information Retrieval (ISMIR) (2011)
5. Bischoff, K., Firan, C.S., Paiu, R., Nejdl, W., Laurier, C., Sordo, M.: Music mood and theme classification – a hybrid approach. In: Proceedings of the 10th International Society for Music Information Retrieval Conference, ISMIR, pp. 657–662 (2009). https://doi.org/10.5281/zenodo.1417317
6. Bischoff, K., Firan, C.S., Paiu, R., Nejdl, W., Laurier, C., Sordo, M.: Music mood and theme classification – a hybrid approach. Poster Session p. 6 (2009)
7. Bogdanov, D., Wack, N., Gómez Gutiérrez, E., Gulati, S., Boyer, H., Mayor, O., Roma Trepat, G., Salamon, J., Zapata González, J.R., Serra, X., et al.: Essentia: an audio analysis library for music information retrieval. In: Dixon, S., Britto, A., Gouyon, F. (eds.) Proceedings of the 14th of the International Society for Music Information Retrieval Conference (ISMIR), ISMIR, pp. 493–498. International Society for Music Information Retrieval (ISMIR) (2013)
8. Bradley, M.M., Lang, P.J.: Affective norms for English words (ANEW): Instruction manual and affective ratings (1999)
9. Cabrera, D., et al.: PsySound: a computer program for psychoacoustical analysis. In: Proceedings of the Australian Acoustical Society Conference, vol. 24, pp. 47–54. AASC Melbourne (1999)
10. Cano, E., Morisio, M.: Moodylyrics: a sentiment annotated lyrics dataset. In: Proceedings of the 2017 International Conference on Intelligent Systems, Metaheuristics and Swarm Intelligence, ISMSI'17, pp. 118–124. Association for Computing Machinery, New York (2017). https://doi.org/10.1145/3059336.3059340

11. Chaki, S., Doshi, P., Patnaik, P., Bhattacharya, S.: Attentive RNNs for continuous-time emotion prediction in music clips. In: Proceedings of the 3rd Workshop on Affective Content Analysis, pp. 36–46. AAAI (2020)
12. Chang, W.H., Li, J.L., Lin, Y.S., Lee, C.C.: A genre-affect relationship network with task-specific uncertainty weighting for recognizing induced emotion in music. In: Proceedings of the 2018 IEEE International Conference on Multimedia and Expo (ICME), pp. 1–6. IEEE (2018). https://doi.org/10.1109/ICME.2018.8486570
13. Chen, Y.A., Wang, J.C., Yang, Y.H., Chen, H.: Linear regression-based adaptation of music emotion recognition models for personalization. In: Proceedings of the IEEE International Conference on Acoustics, Speech and Signal Processing (ICASSP), pp. 2149–2153. IEEE (2014). https://doi.org/10.1109/ICASSP.2014.6853979
14. Chen, Y.A., Yang, Y.H., Wang, J.C., Chen, H.: The amg1608 dataset for music emotion recognition. In: 2015 IEEE International Conference on Acoustics, Speech and Signal Processing (ICASSP), pp. 693–697. IEEE, South Brisbane (2015). https://doi.org/10.1109/ICASSP.2015.7178058
15. Chowdhury, S., Vall, A., Haunschmid, V., Widmer, G.: Towards explainable music emotion recognition: the route via mid-level features. In: Proceedings of the 20th International Society for Music Information Retrieval Conference, ISMIR, pp. 237–243 (2019). arXiv:1907.03572
16. Cross, I.: Music, cognition, culture, and evolution. Ann. N. Y. Acad. Sci. **930**(1), 28–42 (2001). https://doi.org/10.1111/j.1749-6632.2001.tb05723.x
17. Delbouys, R., Hennequin, R., Piccoli, F., Royo-Letelier, J., Moussallam, M.: Music mood detection based on audio and lyrics with deep neural net. In: Proceedings of the 19th International Society for Music Information Retrieval Conference, ISMIR, pp. 370–375 (2018)
18. Devlin, J., Chang, M.W., Lee, K., Toutanova, K.: Bert: pre-training of deep bidirectional transformers for language understanding. In: Proceedings of the 2019 Conference of the North American Chapter of the Association for Computational Linguistics: Human Language Technologies, vol. 1 (Long and Short Papers), pp. 4171–4186 (2019). https://doi.org/10.18653/v1/N19-1423
19. Dong, Y., Yang, X., Zhao, X., Li, J.: Bidirectional convolutional recurrent sparse network (BCRSN): an efficient model for music emotion recognition. IEEE Trans. Multimedia **21**(12), 3150–3163 (2019). https://doi.org/10.1109/TMM.2019.2918739
20. Donnelly, P.J., Beery, A.: Evaluating large-language models for dimensional music emotion prediction from social media discourse. In: Abbas, M., Freihat, A.A. (eds.) Proceedings of the 5th International Conference on Natural Language and Speech Processing (ICNLSP 2022), pp. 242–250. Association for Computational Linguistics (2022)
21. Donnelly, P.J., Gaur, S.: Mood dynamic playlist: interpolating a musical path between emotions using a KNN algorithm. In: Ahram, T., Taiar, R. (eds.) Human Interaction & Emerging Technologies: Artificial Intelligence & Future Applications (IHIET-AI 2022), vol. 23. AHFE Open Access (2022). https://doi.org/10.54941/ahfe100894
22. Eyben, F., Wöllmer, M., Schuller, B.: OpenSMILE: the Munich versatile and fast open-source audio feature extractor. In: Proceedings of the 18th ACM International Conference on Multimedia, MM'10, pp. 1459–1462. Association for Computing Machinery, New York (2010). https://doi.org/10.1145/1873951.1874246
23. Fan, J., Tatar, K., Thorogood, M., Pasquier, P.: Ranking-based emotion recognition for experimental music. In: Proceedings of the 18th International Society for Music Information Retrieval Conference, ISMIR, vol. 2017, pp. 368–375 (2017). https://doi.org/10.5281/zenodo.1416946
24. Hizlisoy, S., Yildirim, S., Tufekci, Z.: Music emotion recognition using convolutional long short term memory deep neural networks. Int. J. Eng. Sci. Technol. **24**(3), 760–767 (2021). https://doi.org/10.1016/j.jestch.2020.10.009
25. Hu, X., Downie, J.S.: Improving mood classification in music digital libraries by combining lyrics and audio. In: Proceedings of the 10th Annual Joint Conference on Digital libraries, JCDL'10, pp. 159–168. Association for Computing Machinery, New York (2010). https://doi.org/10.1145/1816123.1816146

26. Hu, X., Downie, J.S.: When lyrics outperform audio for music mood classification: a feature analysis. In: Proceedings of the 11th International Society for Music Information Retrieval Conference, ISMIR, pp. 619–624 (2010)
27. Hu, X., Downie, J.S., Ehmann, A.F.: Lyric text mining in music mood classification. In: Proceedings of the 10th International Society for Music Information Retrieval Conference, ISMIR, vol. 183, pp. 2–209 (2009). https://doi.org/10.5281/zenodo.1416790
28. Hu, X., Li, F., Ng, T.D.J.: On the relationships between music-induced emotion and physiological signals. In: Proceedings of the 19th International Society for Music Information Retrieval Conference, ISMIR, pp. 362–369 (2018). https://doi.org/10.5281/zenodo.1492425
29. Huang, I.S., Lu, Y.H., Shafiq, M., Ali Laghari, A., Yadav, R.: A generative adversarial network model based on intelligent data analytics for music emotion recognition under IoT. Mob. Inf. Syst. **2021**, 1–8 (2021). https://doi.org/10.1155/2021/3561829
30. Huang, Q., Jansen, A., Lee, J., Ganti, R., Li, J.Y., Ellis, D.P.W.: MuLan: a joint embedding of music audio and natural language. In: Proceedings of the 23rd International Society for Music Information Retrieval Conference, ISMIR, pp. 559–566 (2022). https://doi.org/10.5281/zenodo.7316724
31. Kingma, D.P., Ba, J.: Adam: a method for stochastic optimization. In: International Conference on Learning Representations (2017). https://doi.org/10.48550/arXiv.1412.6980. ArXiv:1412.6980 [cs]
32. Koelsch, S.: Brain correlates of music-evoked emotions. Nat. Rev. Neurosci. **15**(3), 170–180 (2014). https://doi.org/10.1038/nrn3666
33. Koelstra, S., Muhl, C., Soleymani, M., Lee, J.S., Yazdani, A., Ebrahimi, T., Pun, T., Nijholt, A., Patras, I.: DEAP: a database for emotion analysis using physiological signals. IEEE Trans. Affect. Comput. **3**(1), 18–31 (2012). https://doi.org/10.1109/T-AFFC.2011.15
34. Laurier, C., Grivolla, J., Herrera, P.: Multimodal music mood classification using audio and lyrics. In: 2008 7th International Conference on Machine Learning and Applications, pp. 688–693 (2008). https://doi.org/10.1109/ICMLA.2008.96
35. Laurier, C., Sordo, M., Serra, J., Herrera, P.: Music mood representations from social tags. In: Proceedings of the 10th International Society for Music Information Retrieval Conference, ISMIR, pp. 381–386 (2009). https://doi.org/10.5281/zenodo.1415600
36. Li, T., Ogihara, M.: Detecting emotion in music. In: Proceedings of the 4th International Society for Music Information Retrieval Conference, ISMIR, pp. 1–2 (2003). https://doi.org/10.5281/zenodo.1417293
37. Li, X., Tian, J., Xu, M., Ning, Y., Cai, L.: DBLSTM-based multi-scale fusion for dynamic emotion prediction in music. In: 2016 IEEE International Conference on Multimedia and Expo (ICME), pp. 1–6. IEEE (2016). https://doi.org/10.1109/ICME.2016.7552956
38. Lin, T., Wang, Y., Liu, X., Qiu, X.: A survey of transformers. AI Open **3**, 111–132 (2022). https://doi.org/10.1016/j.aiopen.2022.10.001
39. Liu, H., Fang, Y., Huang, Q.: Music emotion recognition using a variant of recurrent neural network. In: Proceedings of the 2018 International Conference on Mathematics, Modeling, Simulation and Statistics Application (MMSSA), pp. 15–18. Atlantis Press (2019). https://doi.org/10.2991/mmssa-18.2019.4
40. Liu, Y., Liu, Y., Zhao, Y., Hua, K.A.: What strikes the strings of your heart?-feature mining for music emotion analysis. IEEE Trans. Affect. Comput. **6**(3), 247–260 (2015). https://doi.org/10.1109/TAFFC.2015.2396151
41. Liu, Y., Ott, M., Goyal, N., Du, J., Joshi, M., Chen, D., Levy, O., Lewis, M., Zettlemoyer, L., Stoyanov, V.: Roberta: a robustly optimized bert pretraining approach (2019). https://doi.org/10.48550/arXiv.1907.11692. ArXiv:1907.11692 [cs]
42. Lu, L., Liu, D., Zhang, H.J.: Automatic mood detection and tracking of music audio signals. IEEE Trans. Audio Speech Lang. Process. **14**(1), 5–18 (2006). https://doi.org/10.1109/TSA.2005.860344
43. Ma, Y., Li, X., Xu, M., Jia, J., Cai, L.: Multi-scale context based attention for dynamic music emotion prediction. In: Proceedings of the 25th ACM International Conference on Multimedia, pp. 1443–1450. ACM (2017). https://doi.org/10.1145/3123266.3123408

44. Malheiro, R., Panda, R., Gomes, P., Paiva, R.P.: Emotionally-relevant features for classification and regression of music lyrics. IEEE Trans. Affect. Comput. **9**(2), 240–254 (2016). https://doi.org/10.1109/TAFFC.2016.2598569
45. Mathieu, B., Essid, S., Fillon, T., Prado, J., Richard, G.: Yaafe, an easy to use and efficient audio feature extraction software. In: Proceedings of the 11th International Society for Music Information Retrieval Conference, ISMIR, vol. 2010, pp. 441–446 (2010). https://doi.org/10.5281/zenodo.1418321
46. McKay, C., Fujinaga, I., Depalle, P.: jAudio: a feature extraction library. In: Proceedings of the 6th International Conference on Music Information Retrieval, ISMIR, pp. 600–603 (2005). https://doi.org/10.5281/zenodo.1416648
47. Orjesek, R., Jarina, R., Chmulik, M., Kuba, M.: DNN based music emotion recognition from raw audio signal. In: 2019 29th International Conference Radioelektronika (RADIOELEKTRONIKA), pp. 1–4. IEEE (2019). https://doi.org/10.1109/RADIOELEK.2019.8733572
48. Panda, R., Malheiro, R.M., Paiva, R.P.: Audio features for music emotion recognition: a survey. IEEE Trans. Affect. Comput. (2020). https://doi.org/10.1109/TAFFC.2020.3032373
49. Radford, A., Wu, J., Child, R., Luan, D., Amodei, D., Sutskever, I.: Language models are unsupervised multitask learners
50. Russell, J.A.: A circumplex model of affect. J. Pers. Soc. Psychol. **39**(6), 1161–1178 (1980). https://doi.org/10.1037/h0077714
51. Sanh, V., Debut, L., Chaumond, J., Wolf, T.: Distilbert, a distilled version of bert: smaller, faster, cheaper and lighter (2019). https://doi.org/10.48550/arXiv.1910.01108. ArXiv:1910.01108 [cs.CL]
52. Shrestha, A., Mahmood, A.: Review of deep learning algorithms and architectures. IEEE Access **7**, 53040–53065 (2019). https://doi.org/10.1109/ACCESS.2019.2912200
53. Soleymani, M., Aljanaki, A., Yang, Y.H., Caro, M.N., Eyben, F., Markov, K., Schuller, B.W., Veltkamp, R., Weninger, F., Wiering, F.: Emotional analysis of music: a comparison of methods. In: Proceedings of the 22nd ACM International Conference on Multimedia, pp. 1161–1164 (2014). https://doi.org/10.1145/2647868.2655019
54. Soleymani, M., Caro, M.N., Schmidt, E.M., Sha, C.Y., Yang, Y.H.: 1000 songs for emotional analysis of music. In: Proceedings of the 2nd ACM International Workshop on Crowdsourcing for Multimedia, CrowdMM'13, pp. 1–6. Association for Computing Machinery, New York (2013). https://doi.org/10.1145/2506364.2506365
55. Tang, Q., Huang, Z., Zhou, H., Ye, P.: Effects of music therapy on depression: a meta-analysis of randomized controlled trials. PLOS ONE **15**(11), 1–23 (2020). https://doi.org/10.1371/journal.pone.0240862
56. Tzanetakis, G., Cook, P.: Marsyas: a framework for audio analysis. Organised Sound **4**(3), 169–175 (2000). https://doi.org/10.1017/S1355771800003071
57. Vaswani, A., Shazeer, N., Parmar, N., Uszkoreit, J., Jones, L., Gomez, A.N., Kaiser, L., Polosukhin, I.: Attention is all you need. In: Proceedings of the 31st International Conference on Neural Information Processing Systems, NIPS'17, pp. 6000–6010. Curran Associates Inc., Red Hook (2017). arxiv.org/abs/1706.03762v5
58. Wang, J.C., Yang, Y.H., Chang, K., Wang, H.M., Jeng, S.K.: Exploring the relationship between categorical and dimensional emotion semantics of music. In: Proceedings of the 2nd International ACM Workshop on Music Information Retrieval with User-Centered and Multimodal Strategies (MIRUM), pp. 63–68. ACM Press, Nara (2012). https://doi.org/10.1145/2390848.2390865
59. Wang, J.C., Yang, Y.H., Chang, K., Wang, H.M., Jeng, S.K.: Exploring the relationship between categorical and dimensional emotion semantics of music, pp. 63–68. ACM, Nara (2012). https://doi.org/10.1145/2390848.2390865
60. Warriner, A.B., Kuperman, V., Brysbaert, M.: Norms of valence, arousal, and dominance for 13,915 English lemmas. Behav. Res. Methods **45**(4), 1191–1207 (2013). https://doi.org/10.3758/s13428-012-0314-x
61. Weninger, F., Eyben, F., Schuller, B.: On-line continuous-time music mood regression with deep recurrent neural networks. In: 2014 IEEE International Conference on Acoustics, Speech

and Signal Processing (ICASSP), pp. 5412–5416. IEEE (2014). https://doi.org/10.1109/ICASSP.2014.6854637
62. Wu, T.L., Jeng, S.K.: Probabilistic estimation of a novel music emotion model. In: Proceedings of the 14th International Conference on Advances in Multimedia Modeling, MMM'08, pp. 487–497. Springer, Berlin/Heidelberg (2008). https://doi.org/10.1007/978-3-540-77409-9_46
63. Yang, D., Lee, W.: Disambiguating music emotion using software agents. In: Proceedings of the 5th Annual Meeting of the International Society for Music Information Retrieval, p. 6 (2004). https://doi.org/10.5281/zenodo.1415271
64. Yang, Y.H., Chen, H.H.: Machine recognition of music emotion: a review. ACM Trans. Intell. Syst. Technol. **3**(3), 1–30 (2012). https://doi.org/10.1145/2168752.2168754
65. Yang, Y.H., Lin, Y.C., Cheng, H.T., Liao, I.B., Ho, Y.C., Chen, H.H.: Toward multimodal music emotion classification. In: Proceedings of the 9th Pacific Rim Conference on Multimedia, pp. 70–79. Springer (2008). https://doi.org/10.1007/978-3-540-89796-5_8
66. Yang, Z., Dai, Z., Yang, Y., Carbonell, J., Salakhutdinov, R., Le, Q.V.: Xlnet: generalized autoregressive pretraining for language understanding (2020). https://doi.org/10.48550/arXiv.1906.08237. ArXiv:1906.08237 [cs]
67. Zhang, K., Zhang, H., Li, S., Yang, C., Sun, L.: The PMEmo dataset for music emotion recognition. In: Proceedings of the 2018 International Conference on Multimedia Retrieval, pp. 135–142. ACM, Yokohama (2018). https://doi.org/10.1145/3206025.3206037
68. Zhang, Y., Jiang, J., Xia, G., Dixon, S.: Interpreting song lyrics with an audio-informed pre-trained language model. In: Proceedings of the 23rd International Society for Music Information Retrieval Conference, pp. 19–26. ISMIR, Bengaluru (2022). https://doi.org/10.5281/zenodo.7316584

Aggregating Industrial Security Findings with Semantic Similarity-Based Techniques

Markus Voggenreiter, Phillip Schneider, and Abdullah Gulraiz

1 Introduction

Automating security tests are a common practice in modern industrial software engineering. Security tools analyze software from various perspectives to achieve an exhaustive picture of the security status of a software product within Continuous Integration or Continuous Deployment (CI/CD) pipelines. These tests output semi-structured reports containing potential security shortcomings, the so-called security findings. While this fosters the principles of modern software development, reduces the testing effort, and provides early and continuous insights into product security, it also comes at a cost.

To avoid white spots in the analysis, overlapping coverages between tools are common in industrial development. In combination with different perspectives allowing the identification of the same finding, the occurrence of duplicates or almost identical findings is inevitable. These duplicates falsify the security overview and challenge security professionals and developers alike in their day-to-day work. Consequently, the identification and elimination of duplicate findings are crucial for efficient software engineering. Considering the overall amount of findings per project and the frequency of new reports being generated, the manual identification of duplicates by the project team is not feasible.

Previously, we investigated the potential of semantic similarity-based clustering techniques to identify and aggregate these duplicate findings. Natural Language

M. Voggenreiter (✉)
Siemens Technology/LMU, Munich, Germany
e-mail: markus.voggenreiter@siemens.com

P. Schneider · A. Gulraiz
Department of Computer Science, Technical University of Munich, Munich, Germany
e-mail: phillip.schneider@tum.de; abdullah.gulraiz@tum.de

M. Abbas (ed.), *Practical Solutions for Diverse Real-World NLP Applications*, Signals and Communication Technology,
https://doi.org/10.1007/978-3-031-44260-5_7

Processing (NLP) has not only shown promising results for clustering texts from medicine, linguistics, and software engineering [1, 3, 7] but also effectively aggregate security findings [11]. This work was continued by validating our results in an industrial context. In this article, we present the outline of our former work and put it into the context of industrial software development projects. We investigate the relevance of our results for industrial practice and evaluate their impact by integrating three different clustering approaches into three distinct industrial projects.

The remainder of this article is structured as follows. Section 2 presents background information on the aggregation of security findings and gives an overview of related work on applying NLP techniques. Section 3 presents our previous research in the domain of security findings aggregation using semantic similarity techniques. The results of our research are discussed in Sect. 4 in the context of industrial practice. The application of our research to industrial practice is demonstrated in Sect. 5. Section 6 concludes the article with a summary and an outlook toward future work.

2 Background and Related Work

This section provides background information on security findings, occurrence of duplicates, and related studies in the domain of NLP techniques.

For the first step in order to understand the challenges of duplicate security findings in software development projects, it must be clarified what a security finding is. The domain of software security contains multiple terms to precisely describe different shortcomings in the security of a software product, including vulnerability, security weakness, security defect, or bug. However, we want to utilize the term *security finding* to summarize all of them and avoid further refinement of the terminology. Consequently, we define a *security finding* as any potential threat to the security of a product that was found but not yet verified or further processed.

A software engineering project might now be confronted with duplicate findings if the exact same problem with the software is described multiple times. This can occur, e.g., if multiple tools have overlapping software coverage and consequently identify the same finding twice, resulting in a duplicate entry in the final list of security findings (inter-source duplicate). However, also the same source might identify a problem at different locations (intra-source duplicate). According to our scope, both represent duplicates and therefore necessitate correction. Besides this problem-based perspective on duplicates, other strategies exist that consider the finding's location or its underlying solution as well. For this article, we focus on the problem-based perspective.

Furthermore, it is necessary to elaborate on the activities that generate security reports with duplicate findings. Security testing can be categorized according to multiple properties depending on the testing strategy, involved testers, tested components, and numerous others. We limit our categorization to those security tests that can be automated in pipelines and scan an actual part of the product. Further, we categorize them into two major categories: tests that examine the static elements of the software (e.g., code, configuration, or dependencies) are called static application security testing (SAST), and tests performed against the dynamic, actually running application are called dynamic application security testing (DAST). This separation represents a clear distinction, as static testing can only guess whether a finding is actually affecting the software, while dynamic techniques directly identify the exploitable security finding.

From our analysis of the literature on security findings management, we found that there are no NLP-related publications focusing on the identification of duplicate security findings. However, a number of NLP methods have been successfully applied to related subdomains in the software engineering field. For example, Kuhn et al. [5] use latent semantic indexing (*LSI*) and clustering to analyze linguistic information found in source code, such as identifier names or comments, to reveal topics and support program comprehension. In a study from [10], a corpus of app reviews with comments about a variety of software issues is clustered into topics with problem-specific issue categories. Another study from [4] focuses on automatically forming semantic clusters of functional requirements based on cosine similarity with a corpus of documents containing software requirements specifications. The authors conduct an empirical evaluation of agglomerative hierarchical clustering using four open-access software projects. In order to assess the software quality of programs, [12] apply a hierarchical cluster algorithm to create problem-oriented clusters, reducing the effort needed to review the code. The study shows that semantic clusters are an effective technique for defect prediction.

3 Investigation of Semantic Similarity-Based Techniques

In this section, we summarize our previous work on the semantic similarity-based clustering of security findings, published in Schneider et al. [11]. In the previous work, we analyzed the potential of semantic similarity-based clustering for the clustering of security findings.

3.1 Preparation Phase

To quantify the performance of different semantic similarity techniques, a ground-truth benchmark dataset is required, enabling the comparison between human-labeled clusters and predictions of the semantic similarity algorithms. Therefore, we asked two security professionals from the industry to annotate semantically duplicate findings in a given list of security reports. Due to the significant differences in perspective between SAST and DAST reports, we decided to construct two separate datasets, each of which comprising reports from only one testing type.

A significant challenge in constructing such a dataset is the content of the security tool reports. Security tool reports are often exported as JSON files containing security finding objects. Across different tools, these reports utilize different schemas, resulting in different property names referring to the same finding feature (e.g., *description*, *FullDescription*, *text*, *Message*, or *details*). For the construction, the security professionals consolidate semantically duplicate findings from all tool reports of a testing iteration based on certain features, e.g., description, location, or unique identifier. Therefore, they need to find the feature in the respective tool schema and compare it to the other findings. Manually annotating such a dataset would require them to memorize $N \times M$ property names when identifying N features across M distinct security testing reports. To enhance efficiency and reduce manual, repetitive work, we introduced the Security Findings Labeler (*SeFiLa*).[1] This tool allows security professionals to upload reports from different security tools and conveniently group all findings into named clusters.

The initial, unconsolidated reports of the dataset were generated by scanning the open-source, vulnerable web application JuiceShop[2] with seven SAST tools and two DAST tools. For reproducibility reasons, we solely selected tools free of charge that can be reasonably automated in real-world software development pipelines. Following the categorization of SAST tools by Angermeir et al. [2], we looked into three third-party vulnerability scanner (Anchore, Dependency Check, Trivy), three static code analysis tools (HorusSec, Semgrep, CodeQL), and one secret detection tool (Gitleaks). For DAST tools we selected two web application scanners (Arachni, OWASP ZAP). From each tool, one report was taken for the dataset. The security professionals assigned findings to named clusters representing the same security problem. This process was aided by features like the CVE-ID (common vulnerabilities and exposures) which provides an identifier and a reference method for publicly known security vulnerabilities. Other helpful features are descriptions and solutions generated by the testing tools. After all, findings were assigned to clusters, the dataset comprising our baseline for duplicate identification was completed. The dataset and the code to run the test cases were published in a public GitHub repository.[3]

3.2 Dataset Description

After labeling the exported security findings with our annotation tool *SeFiLa*, the professionals provided us with one dataset for the SAST and DAST findings each. The descriptive statistics of both datasets are summarized in Table 1. We observe that SAST findings are far more frequent, making up 97.4% of the total

[1] https://github.com/abdullahgulraiz/SeFiLa

[2] https://owasp.org/www-project-juice-shop/

[3] https://github.com/abdullahgulraiz/SeFiDeF

Table 1 Data records from static analysis security tools (SAST) and dynamic analysis security tools (DAST)

Statistic	SAST	DAST
Number of clusters	183	10
Number of findings	1351	36
Avg. findings per cluster	7	3
Avg. characters per finding	302	471
Min. findings per cluster	1	1
Max. findings per cluster	408	25

findings. The number of formed clusters for the SAST findings is significantly higher than for DAST findings. While both datasets had clusters with only one finding, the maximum cluster size was by far larger in the SAST dataset. Despite these discrepancies, the average number of findings per cluster is not too different between the datasets, ranging from a mean value of 3 for DAST to a mean value of 7 for SAST findings. In addition, DAST finding texts are more verbose since they contain 169 more characters on average.

To investigate the potential of semantic similarity techniques, constructing the finding string from the finding features is crucial. Analyzing the initial dataset, we identified that solely a single feature describing the finding is consistently found across all SAST tools. For the DAST findings, multiple features, including the description, a name, and even a solution/mitigation, were consistently found across all findings. Furthermore, we observed that DAST features are sufficiently verbose to comprehend the problem from their finding string and thereby contain enough semantic content for semantic clustering. Contrarily, we find SAST features to be very brief, for that matter, making it almost impossible to understand a finding just from the single consistently existing feature.

3.3 Analysis Phase

To analyze the potential of semantic similarity-based methods, we investigated techniques commonly proposed in the literature. We decided for three popular techniques often used as baseline models: knowledge graph-based similarity with *WordNet* [8], *LSI* [6], and *SBERT* [9].

To compare these models and test their potential, we provided the same problem-specific finding strings to each one and used them to determine similar findings. These strings were constructed by extracting all findings from the security reports and concatenating selected features of each finding. To counteract the limitation of very short SAST finding strings, we made use of CVE-IDs to increase the textual content of SAST finding strings. By leveraging the CVE identifier present in some findings, we concatenated finding strings of various machine-generated descriptions with the same CVE-ID. This allows for more semantic content and longer descriptions of the underlying problem. The verbosity of DAST findings on the other hand introduced the opportunity to check whether more features lead to better results. This step led us to construct a total of four corpora with finding strings from both SAST and DAST datasets for the identification of duplicate findings, as listed below:

- ***SAST-D*:** consists only of SAST finding descriptions
- ***SAST-ConcD*:** consists of concatenated SAST finding descriptions with the same CVE-ID
- ***DAST-NDS*:** consists of concatenated DAST finding names, descriptions, and solution texts
- ***DAST-D*:** consists only of DAST finding descriptions

Our methods express the similarity between two findings as a score between 0 and 1 where 1 indicates the highest similarity. Hence, we established a *similarity threshold* for each experiment defining the value above which two finding strings are deemed to be semantically similar. The respective findings belonging to similar finding strings are grouped to form predicted clusters.

In advance of comparing the predicted clusters to our ground-truth dataset, we assessed the logical soundness of the predicted clusters. In practice, each finding belongs to exactly one cluster, implying transitivity of the clustering. However, in certain cases the problem description of two findings was identical, but it was repeated in one finding for multiple instances, leading to a discrepancy in text length. Since the similarity depends on the similarity of the finding strings, we encounter the following example predictions with *Similar Findings* listed in descending order of semantic similarity scores with the corresponding *Finding* identifier:

$$\{Finding : 1,\ Similar\ Findings : \{1, 2, 4\}\}$$

$$\{Finding : 2,\ Similar\ Findings : \{2, 1, 3, 5\}\}$$

Let us assume that findings $\{1, 2, 3\}$ contain the same problem description, although it appears once in $Finding\,1$, two times in $Finding\,2$, and three times in $Finding\,3$. While $Finding$ 1 is found similar to findings $\{1, 2, 4\}$, its similarity score with respect to $Finding$ 3 is below the clustering threshold due to the varying text length. However, $Finding$ 2 does have $Finding$ 3 in its set of similar findings. If $Finding$ 3 is similar to $Finding$ 2, it should also be similar to $Finding$ 1, regardless of repetitive text. Therefore, even though $Finding\,3$ exists only in the set of similar findings for $Finding\,2$, it should appear in the final set of similar findings of $Finding$ 1 as well. In our initial clustering experiments and discussions with the security professional, we observed that while lowering the similarity threshold led to many false positive predictions, transitive clustering improved the results without changing the similarity threshold. Therefore, we refined all three methods by applying the transitive property after they finished their predictions. This change in the processing causes the above predictions to be adapted to

$$\{Finding : 1,\ Similar\ Findings : \{1, 2, 3, 4, 5\}\}$$

$$\{Finding : 2,\ Similar\ Findings : \{1, 2, 3, 4, 5\}\}$$

After transitive clustering, we removed the duplicate clusters from predictions and compared the final predictions against the ground-truth clusters.

In addition to the quantitative evaluation with performance measures, we collected qualitative feedback from the security professionals on incorrectly clustered findings. We limited the information about each finding to the finding strings used as input for the NLP techniques and asked for possible reasons for the incorrect clustering. This created a list of reasons that led to poor duplicate identification from the perspective of a domain-aware security professional. Finally, each incorrect cluster is associated with at least one reason for the incorrect clustering, providing insights into the different challenges and their prevalence in the results. The evaluation was aided by *SeFiLa* for annotation of the findings, assignment of reasons, and documentation.

3.4 Evaluation Results

The summary of the quantitative results achieved when applying semantic clustering using a technique from each category of semantic similarity methods to each of the four corpora is presented in Table 2. The experiments were performed for similarity thresholds 0.1 $\leq$ and $\leq$ 0.95. The performance metric values for the experiment with the highest F-score are reported. For the SAST findings, the highest F-score of 0.857 was achieved by applying *LSI* to the *SAST-ConcD* corpus. For DAST, the F-score of 0.857 was achieved by utilizing *SBERT* on the more verbose *DAST-NDS* corpus.

Figures 1 and 2 show the F-scores of different technique-corpus combinations over different similarity thresholds for SAST and DAST, respectively. We see that

Table 2 Summary table of performance metrics (highlighted results show the best performing techniques for SAST and DAST)

Technique	Corpus	Metrics		
		F-score	Precision	Recall
SBERT	*SAST-D*	0.709	0.621	0.825
	SAST-ConcD	0.797	0.701	0.923
	DAST-NDS	**0.857**	**0.818**	**0.900**
	DAST-D	0.857	0.818	0.900
LSI	*SAST-D*	0.739	0.658	0.842
	SAST-ConcD	**0.816**	**0.734**	**0.918**
	DAST-NDS	0.857	0.818	0.900
	DAST-D	0.857	0.818	0.900
KG	*SAST-D*	0.659	0.556	0.809
	SAST-ConcD	0.777	0.676	0.913
	DAST-NDS	0.727	0.667	0.800
	DAST-D	0.727	0.667	0.800

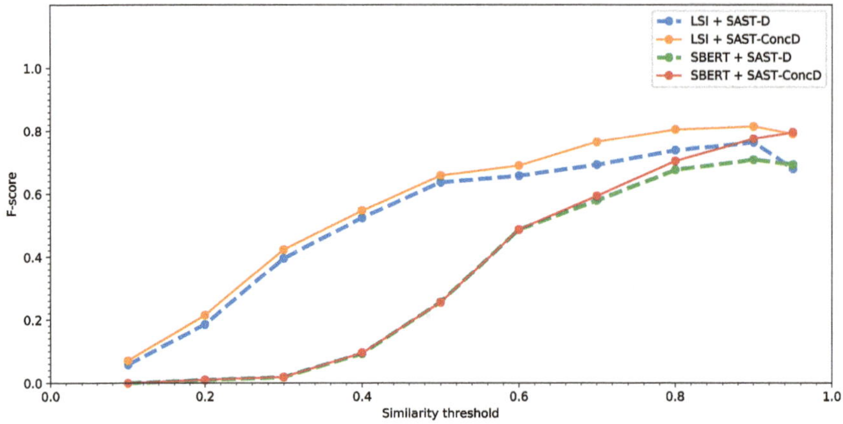

Fig. 1 Semantic clustering results of SAST findings for different similarity thresholds

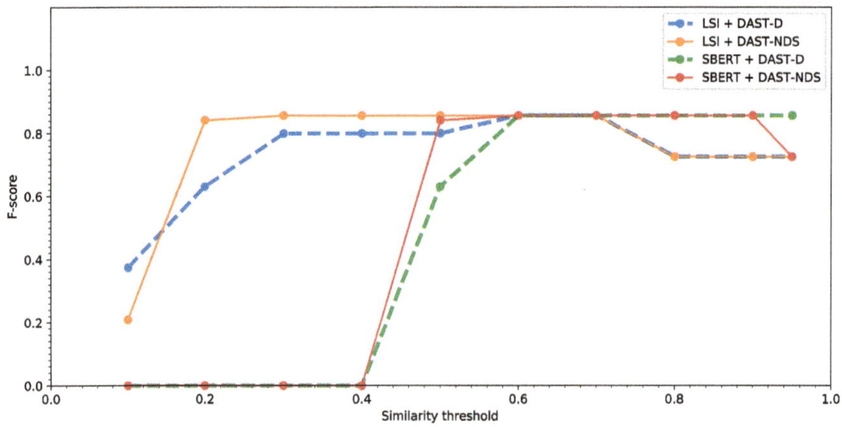

Fig. 2 Semantic clustering results of DAST findings for different similarity thresholds

the F-scores increase with increasing similarity threshold, peaking at a threshold value ≥ 0.6 for DAST and at around 0.9 for SAST. As the results for the knowledge graph-based clustering yield lower F-scores for all corpora, they are disregarded for further presentation.

For the qualitative evaluation, we showed incorrect predictions from the best results of semantic clustering of SAST and DAST findings to a security professional. The cluster results came from applying *LSI* to *SAST-ConcD* corpus for the SAST dataset and applying *SBERT* to *DAST-NDS* corpus for the DAST dataset. Using *SeFiLa*, the security professional inspected incorrect predictions and their associated ground-truth cluster. The security professional assigned possible reasons for poor duplicate identification by reading finding strings associated with incorrect predictions. These reasons are documented for 72 incorrect SAST predictions and

Table 3 Overview of provided explanations from the qualitative evaluation

Reason	Explanation for incorrect clustering	SAST	DAST
1	In the context of the product, this result can only be identified by somebody knowing the context of the application	–	2
2	Different tools use a different phrasing to explain the same issue	5	–
3	The tools sometimes provide no description of the finding. Hence, the features could only rely on the title	39	–
4	Some tools provide more, and some tools provide less text in their description, which reduces the impact of actual relevant features	19	–
5	Additional review necessary due to an unknown reason for the decision	5	–
6	The sub-optimally constructed feature string could be the reason for the incorrect clustering	39	–
7	The tool describes the finding precisely according to the location of occurrence. Hence the finding text is over-specified	3	–
8	Human annotation error and the suggested clustering by the algorithm are correct	1	–
9	One tool addresses the issue of using an *eval* function, while the other one has the problem of user-controlled values in it. However, it would not be considered as a major false positive	3	–

2 incorrect DAST predictions. The reasons and the number of times they were assigned to an incorrect prediction from either SAST or DAST clusters are listed in Table 3.

4 Discussions of the Preliminary Results

In our previous work, we explored the potential of semantic clustering of security findings through various similarity techniques. We tested three techniques from neural network-based, corpus-based, and knowledge-based methods on carefully constructed identification strings that describe security findings. We found that findings from SAST tools are best clustered by preprocessing their finding string based on CVE-IDs and utilize *LSI* as similarity technique. For findings from DAST tools, we concluded that *SBERT* is recommendable with a finding string that utilizes a higher textual content.

Furthermore, we identified the data quality as a key challenge for the aggregation of security findings with semantic clustering in practice. The finding string used to identify duplicates is constructed by extracting features from each finding. The quality of each feature dictates the correctness of the resulting deduplication. While some tools provide a valid description of the string construction, others solely provide a vague title. Also, the data size encapsulated by these features impacts the deduplication precision. The information size ranges from contextually rich problem descriptions to under-specified ones, leading to different extents of semantic content being captured and, thereby, incorrect predictions being made. To improve SAST findings clustering results, it is evident that the semantic content of finding strings representing a problem must be improved. As discussed in our original work, its results are limited by the degree of realism the security reports generated by testing JuiceShop impose and the external validity that can be provided by a subjective task like deduplication. Consequently, we identified the necessity to apply our results in real-world DevOps scenarios, studying the relevance of our research in practice. Consequently, an integration into industrial projects is crucial.

As identified by Angermeir et al. [2], industrial software development projects use a broad variety of static security testing tools. However, the amount of dynamic tools employed is limited, so that an applied aggregation approach should be optimized for SAST finding. Since *LSI* not only outperformed other methods in the duplicate identification of SAST finding but also showed good performance for DAST findings, it is the method of choice for further investigation. Moreover, the identified necessity of preprocessing SAST findings with their CVE-ID and more context data as well as the importance of postprocessing the results to fulfill the property of transitivity indicate the relevance of further processing stages enclosing the similarity techniques in an application. Consequently, the investigation in industrial practice requires additional effort to achieve valuable insights.

5 Application in Industry

Managing security findings in the industry has to cope with a variety of challenges ranging from ensuring a consistent level of data quality to the reliable prioritization of actions according to the projects demand, as discussed by Voggenreiter and Schöpp [13]. Coping with duplicate findings throughout continuously added security reports is one of them. Our results provide us with a first indication of how this aggregation can be realized. To deepen this knowledge and validate its impact, we integrated the results acquired in our last research into industrial projects.

5.1 Constraints in Practice

The insights achieved by our last research can, however, not be directly transformed into industrial practice. Instead, the aggregation of security findings is affected by three constraints that must be covered before it can benefit actual projects.

As discussed in Sect. 4, the automation of findings clustering as the processing stage is essential. Hence, the similarity technique *LSI* must be embedded in an application, which cares for the preprocessing of findings as well as the postprocessing of clusters to ensure transitivity.

Our initial research found that the similarity technique is still prone to errors, especially if the semantic content comprised by the features differentiates significantly. Consequently, an approach to correct clusters is essential. This necessary information underlying the correction can only be provided by users analyzing the findings and their context. Hence, it is crucial that the clusters can be corrected by user input. As seen in the reasons for incorrect clustering, a "result can only be identified by somebody knowing the context of the application" under certain circumstances. This implies that even a perfect model might not be able to identify all clusters correctly due to low data quality or knowledge dependencies in the software context. Exemplary for this circumstance would be a security tool monitoring the application, which identifies the possibility of an open, administrative network port. The root cause for this port could be an incorrectly hardened system image, where unnecessary software is still running. This shortcoming could also be identified by another security tool checking system hardening measures. These two findings would be considered a duplicate by practitioners as they refer to the same fundamental problem. The hardening tool would be able to pinpoint the finding to a certain software in the system including a detailed description of the software and why it should not be running. The perspective of the monitoring tool, however, solely informs about the open port. It cannot provide any information about the software running behind it or the root cause of incorrect hardening measures. Consequently, there is no indication of a duplicate in the raw information. This can only be given by practitioners knowing the context of the application and the dependencies between open ports and running software.

The last constraint affecting our integration is the necessity for continuous maintenance of the clusters. The DevOps principle of "continuous everything" affects security testing as well. Consequently, new reports containing unknown and known findings alike will continuously arise. The frequency can range from a security report being produced during every commit to more sophisticated tests providing findings at each delivery cycle. To perform a precise and resource-aware clustering, incremental maintenance of the clusters is essential.

5.2 Deduplication Techniques

When managing security findings in industrial software development projects throughout a continuously rising amount of reports, the aggregation of identical findings is crucial. With each new report the exact same, already known findings might appear again identifiable by hashing the finding content (hash-based deduplication). The deduplication of those findings is trivial and imposes almost no False Positives but has a major impact on the overall amount of security findings. However, this represents no deduplication based on semantic features, resulting in a comparably low recall. On the other hand more complex techniques like the *LSI*-based deduplication take semantics into consideration, providing a better recall but also a lower precision in comparison. This generally represents an extension of the hash-based deduplication and has a more significant impact on the data.

To cope with the varying tasks that have to be performed on the data and the respective demand toward precision and recall, we decided to propose three different levels of deduplication.

Level 1 The first level utilizes the naive *hash-based deduplication*, where we expect a high precision. Since this level only supports intra-source deduplication and fails to identify similar findings when the source data is changed (e.g., tool vendor adds new data), we expect a low recall.

Level 2 For the second level, we follow the recommendation of our previous research by preprocessing the findings according to CVE-IDs. In addition to applying the *hash-based deduplication*, this stage additionally deduplicates findings using their unique vulnerability identifier. We define this type of processing as *identifier-based deduplication*. In contrast to the first stage, this level lowers the expected precision, as inaccurate CVE-IDs provided by the security tools result in False Positives. However, it also increases our predicted recall, since findings are deduplicated according to a second criterion. Nevertheless, not every finding has a unique vulnerability identifier or the respective security tool does not provide it in their reports. Hence further potential to achieve a higher recall and, consequently, fewer duplicates exists.

Level 3 This third stage implements the techniques of our previous research by utilizing semantic similarity-based methods to deduplicate security findings. Following the recommended procedure to construct the knowledge-dense finding strings, we utilize the clusters formed by level 2 and concatenate the feature content of every finding within one cluster. In this last level, we expect the highest recall of all levels accompanied by the lowest precision. The implementation of all three levels is discussed in the next paragraphs.

5.3 Implementation

To implement our approach, we followed the idea by Voggenreiter and Schöpp [13] of a semantic knowledge base for the management of industrial security findings. Consequently, our aggregation represents a logical inference based on the information existing in the knowledge base. Figure 3 shows the processing

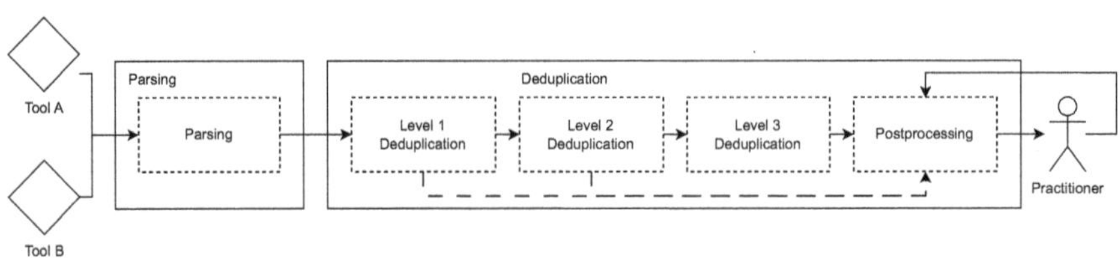

Fig. 3 Implementation inference design

pipeline for the aggregation. In advance to the actual clustering, all reports from the security tools are collected and parsed to a common data format. This abstracts the necessity to identify which field within the reports contains what feature into a separate inference stage. Furthermore, this inference extracts multiple findings from each security report.

After the preparation of the findings data, the actual deduplication is taking place, represented by the *Deduplication* inference in Fig. 3. Each parsed security finding is aggregated with the existing list of security findings using the *hash-based deduplication*. Consequently, a hash value is constructed for each finding that comprises its entire content, excluding the timestamp of the report, the tool version used to identify the finding, and the location, where the finding exists. This exclusion is essential, as the fields either introduce an avoidable negative impact on the recall or dissect the exact same finding at different locations. Minding that new security reports will be continuously added, this processing considers the existing findings as well as the newly added ones. Depending on the demand for precision and recall, these clusters of findings are directly forwarded to the postprocessing or further levels of deduplication are applied. If the *identifier-based deduplication* should also be applied, all findings containing the same unique identifier (in our implementation the CVE-ID) are grouped. Each group represents a unique finding according to its CVE-ID. Afterwards, the third level of deduplication can be applied by constructing a finding string for each group identified during the second level. This string consists of the concatenated *Title* and *Description* field of every finding. For the implementation of *LSI* we kept the configuration and code of the initially evaluated approach and set the similarity threshold to the recommended 0.9 for *LSI* with static tools. Finally, this level provides us with groups of findings that have been deduplicated by all three defined levels. Solely the transitive property of the groups must be controlled and potentially corrected.

These final checks as well as other constraints are implemented in the post-processing. First, the postprocessing ensures that if findings occur in multiple groups after applying the *semantic similarity-based deduplication*, these groups are summarized and duplicate groups eliminated as described in Sect. 3.3. Next, these refined groups are corrected by any existing user input. This input might remove elements from groups or merge two groups. Each group is investigated for applicable user input and the resulting action applied. If the user input conflicts with each other, the separation of groups is prioritized to allow a manual correction of the precision. Finally, each security tool provides different content for certain data fields. Hence, the grouping results in multiple pieces of information being available for the same field. In order to aggregate all duplicates into a single finding, the data fields must be aggregated as well. To avoid data loss, we decided to transform the data fields to lists containing the content of all grouped findings. Afterwards, the aggregated findings are presented to the practitioners through a web interface.

These practitioners might identify corrections in the deduplication resulting in new user input being added to our knowledge base. The fundamental properties of a knowledge base require that this new knowledge is processed to avoid inconsistencies. Hence, the user input triggers another execution of the *Deduplication* inference.

5.4 Evaluation Planning

Our motivation for applying semantic similarity-based aggregation in practice is the identification of its potential for industrial projects in the software engineering domain. This objective is measured by analyzing the performance of the selected method in terms of correctness and usability. This results in two questions our evaluation seeks to answer.

- **RQ1 –** How correct is the clustering of security findings produced by our implementation of *LSI* in industrial practice?
- **RQ2 –** How useful is our implementation of *LSI* in industrial practice?

Similar to our previous research, the first question explores the correctness of the resulting clusters in terms of precision and recall. The second question represents the subjective usefulness of the approach for practitioners, representing a crucial indicator of the potential of our approach.

To evaluate the potential of semantic similarity-based aggregation of security findings in industry, we applied our implementation to three projects with different scopes for security practitioners. Fixing identified findings is not the only task utilizing data from security testing tools. The knowledge transported by security findings affects several processes and tasks in the software development lifecycle. These range from providing data to other project-external activities like "Top N Bugs" lists to processes with direct project impact like refining the testing strategy by Quality Assurance. All these use cases require a dataset of high quality to work efficiently, necessitating the prevention of duplicate findings. Consequently, we decided that our evaluation shall be tested in three different contexts, covering varying aspects of the software development lifecycle.

The first context in which we investigate the potential of our approach is a security code review. As part of these reviews, the code is typically tested with security code analysis tools, checking for flaws that can be identified automatically before going into more detail with a manual analysis. To ensure a broad coverage of the analyzed code, overlaps in the static code analysis tools are common. In our case, the project checks existing infrastructure code with four tools for infrastructure code security testing (*KICS*, *tfsec*, *terrascan*, and *checkov*) and provides the results to security professionals for review.

The second project represents a traditional software development project employing our implementation as a management system for security findings over the course of six months. Its security testing pipeline covers six tools ranging from secret scanning (*Gitleaks*) over static code analysis for infrastructure and software code (*tfsec*, *Semgrep*, and *Bandit*) to third-party dependency and container scanning

(*Trivy* and *NPM Audit*). The results are forwarded to the development team for security review and improvements.

The third context is a collection of custom-made security testing tools that utilize our implementation as a processing and presentation mechanism. Similar to a traditional development project, multiple tools with overlapping coverage produce reports, which must be aggregated to avoid duplicates. However, the audience in this project is solely security professionals instead of a software development team.

For each context, we followed a strict protocol to evaluate the potential of our *LSI*-based approach for the security findings aggregation. We integrated our implementation for each of the three projects and applied all three deduplication levels subsequently. Consequently, three interview rounds were conducted in each project, summarizing nine qualitative data points. Moreover, we acquired quantitative data on the processing statistics to provide context to the qualitative data. In order to answer the research questions, we analyzed the correctness of the results for each combination of context and deduplication level and interviewed the audience about the results on its usability. Therefore, each project team was asked to review the correctness of the deduplication and its perceived usefulness to the team and provide the results as answers in the interview. Our approach was integrated while each project was already running to achieve realistic data and ensure a valid dataset. Since we required insights for each combination of project and deduplication level, the level was changed after each interview. The knowledge base properties allowed us to entirely rebuild it whenever the level was changed, ensuring the new approach was applied.

5.5 *Evaluation Results*

Our evaluation provided us with quantitative and qualitative data about the success of *LSI* for aggregating security findings in practice. In this section, we present both data sets for each project. Detailed information about the security findings or projects were omitted in accordance with the collaboration agreement with our industry partner.

The first project, dealing with security code review, provided one report from four different tools resulting in 293 parsed findings. Our aggregation level 1 and 2 deduplication resulted in 134 findings taking 12 seconds to compute. The level 3 technique using *LSI* took 14 seconds to reduce the number of findings to 56. This data can be found in Table 4 under "Project 1." During the interviews, the first project team stated the usefulness of the findings aggregation provided by the first and second levels. While they found no False Positives in both results, they mentioned multiple False Negatives, hence findings that were incorrectly not clustered. According to the team, this improved significantly when applying the *LSI*. While they still were not able to identify False Positives, the amount of False Negatives was reduced. However, they were still able to identify False Negatives. For example, they stated the correct clustering of two findings from *Checkov* and

Table 4 Quantitative context data

Data type	Project 1			Project 2			Project 3		
	L1	L2	L3	L1	L2	L3	L1	L2	L3
Number of unique report sources		4			6			3	
Number of reports per source		1			40–110			1	
Number of parsed findings		293			309			2713	
Number of aggregated findings	134	134	56	129	82	73	413	413	413
Computation time (in seconds)	12	12	14	33	33	30	21	21	26

TFSec, addressing the need to enable versioning in AWS S3 buckets. This finding was also identified by *Kics* and *Terrascan* separately but not aggregated, resulting in three finding clusters, where one would be correct. Furthermore, the perceived usefulness was stated as "highly useful without doubt" and the wish to not "[...] miss it anymore in the context of a security code review."

The second project, developing software in an industrial context, provided between 40 and 110 reports for each of the six different tools resulting in 309 parsed findings. Each parsed finding represents its occurrence in all reports, so that a new report adding a known finding does not result in an additional parsed finding. The hash-based deduplication reduced this number to 129 aggregated findings, taking 33 seconds. The same time was required by our level 2 deduplication to construct 82 finding clusters. Finally, the *LSI*-based technique reduced this number to 73 findings in 30 seconds. This data can be found in Table 4 under "Project 2." For the *hash-based* and *identifier-based* deduplication, the project team did not find any False Positives. However, it identified multiple False Negatives during the *hash-based* technique, which were diminished by the *identifier-based* deduplication. The team mentioned the *Trivy* finding "curl: HSTS bypass via IDN" as an example for this impact. This finding occurred three times when applying Level 1 deduplication, even though being the exact same finding. This is due to changing data being provided by the security tool *Trivy* (Description being changed, additional data being added), resulting in different hash values. However, all three contain the same CVE-ID allowing the Level 2 deduplication to provide a better aggregation of just one single finding. This consistent precision, in combination with an increased recall, resulted in high usability, according to the project team. Furthermore, they stated that this level of deduplication is valuable for continuous usage during development. With the usage of *LSI*, however, the team was able to identify two incorrectly clustered findings that had to be manually corrected. These mistakes were also represented in the usability, as the additional effort to correct the findings made the technique less usable in comparison to the *identifier-based* deduplication.

In the third project, our approach was utilized to aggregate findings from three custom-developed tools. Our techniques received one report per tool resulting in 2713 parsed findings. All three levels of deduplication identified 413 finding clusters in 21 to 26 seconds. This data can be found in Table 4 under "Project 3." The team considered all three approaches as equally correct and useful. While they considered the usage of a semantic similarity-based technique for the aggregation of security findings as interesting, they identified no impact on the usability and consequently preferred the *hash-based* approach.

5.6 Discussion

Finally, we want to discuss the results of our evaluation in industry and argue about its potential for industrial projects.

The first obvious result from our approach is the increased computation time necessary when applying *LSI* to deduplicate findings. However, this is not due to poor performance of the selected *LSI* technique, but due to the overall implementation in a semantic knowledge base. In order to apply the technique, the title and description of each finding must be loaded to construct the respective finding string. This increases the required time in comparison to simple queries as performed by *hash-based* or *identifier-based* deduplication. Consequently, a different implementation will likely result in different computation times. However, also the teams did not mention any time-related issues.

Another result from our evaluation is the varying relevance of aggregation techniques based on the use case and its source data. The data from the first project team showed a significant impact on the number of findings and consequently the recall. This can be explained by the overlapping coverage of the report sources, focusing on infrastructure code security with similar terminology in the title and description. Due to the use case itself, requiring that each finding is analyzed, any potential False Positive will be identified and corrected, reducing their negative impact. Moreover, these sources provided no unique CVE-ID, resulting in no impact of the *identifier-based* deduplication. The exact opposite effect was observable in project three. Here neither the *identifier-based* deduplication nor the *LSI* technique could affect the number of findings, since the tools were highly customized to their applicable use case. Consequently, neither CVE-IDs were provided nor were titles or descriptions between findings similar. In particular, the tools were developed and configured in a way that no inter-tool duplicates arose. Hence, the usage of sophisticated techniques was unnecessary for this project. Finally, the results gathered during the second project reproduced the results of our initial research conclusion. While using *LSI* had the most significant impact on recall, it also introduced False Positives in contrast to the other two deduplication levels. Since the tools had less overlap in terminology and coverage than the first project, its impact on the recall was also less significant. The use case represented by this project required a high precision, as not every finding can be analyzed by the development team. Hence, a falsely clustered finding might never be identified, making the wish of the team to return to the *identifier-based* deduplication understandable. Consequently, the *LSI* technique not only showed success in this context but also introduced drawbacks.

Based on this evaluation, we conclude that depending on the area of application, different strategies to aggregate security findings are necessary. Consequently, the answer to our first research question is that *LSI* is mostly correct, and we solely identified two False Positives throughout all three projects applying it. Toward the second question we saw that all teams perceived the *LSI*-based deduplication as useful. However, they preferred other deduplication approaches for their higher precision if crucial for the use case. Nevertheless, the potential of a *LSI*-based approach identified in our last research was reinforced by our evaluation and showed to be beneficial for two of the selected use cases.

This already indicates some limitations of our research. In the evaluation, we only considered a subset of all tasks performed by a security practitioner that could be supported by our technology. Hence, we have not yet fully investigated the potential of semantic similarity-based aggregation. However, we have shown its potential for the most important one: industrial software development projects. Furthermore, the construction of finding strings represents another limitation of our research. A more sophisticated construction approach might yield better results in terms of precision and recall. Moreover, the correctness of our approach was assessed by the project team due to the subjectiveness of findings deduplication. However, this introduces the threat that the teams missed incorrectly clustered findings.

6 Conclusions and Future Work

More than 50% of all security findings were correctly identified as duplicates during our evaluation of clustering techniques in industrial practice. This reinforces the necessity for an automated approach to aggregate security findings in the industry. In this article, we investigated the potential of our previous research results by applying them to three different industrial projects and analyzed the correctness of results and usability for the project team. Our results indicate that the applied *LSI*-based technique can be beneficial for industrial projects. However, they also demonstrate the necessity of customizing the aggregation strategy to each use case.

In the future, we want to continue the research on semantic similarity-based techniques and identify approaches to customize them according to the available source data and applicable use cases.

Acknowledgments The authors want to thank the project teams for supporting this research by evaluating the aggregation results and taking the time for the interviews.

References

1. Aggarwal, K., Timbers, F., Rutgers, T., Hindle, A., Stroulia, E., Greiner, R.: Detecting duplicate bug reports with software engineering domain knowledge. J. Softw.: Evol. Process **29**(3), e1821 (2017). https://onlinelibrary.wiley.com/toc/20477481/2017/29/3

2. Angermeir, F., Voggenreiter, M., Moyón, F., Mendez, D.: Enterprise-driven open source software: a case study on security automation. In: 2021 IEEE/ACM 43rd International Conference on Software Engineering: Software Engineering in Practice (ICSE-SEIP), pp. 278–287 (2021)
3. Demner-Fushman, D., Lin, J.: Answer extraction, semantic clustering, and extractive summarization for clinical question answering. In: Proceedings of the 21st International Conference on Computational Linguistics and 44th Annual Meeting of the Association for Computational Linguistics, pp. 841–848. Association for Computational Linguistics, Sydney (2006)
4. Eyal Salman, H., Hammad, M., Seriai, A.D., Al-Sbou, A.: Semantic clustering of functional requirements using agglomerative hierarchical clustering. Information **9**(9), 222–238 (2018)
5. Kuhn, A., Ducasse, S., Gīrba, T.: Semantic clustering: identifying topics in source code. Inf. Softw. Technol. **49**(3), 230–243 (2007), 12th Working Conference on Reverse Engineering
6. Landauer, T.K., Dumais, S.T. (1997) A solution to Plato's problem: the latent semantic analysis theory of acquisition, induction, and representation of knowledge. Psychol. Rev. **104**, 211–240
7. Majewska, O., McCarthy, D., Vulić, I., Korhonen, A.: Acquiring verb classes through bottom-up semantic verb clustering. In: Proceedings of the Eleventh International Conference on Language Resources and Evaluation (LREC 2018). European Language Resources Association (ELRA), Miyazaki (2018)
8. Miller, G.A.: WordNet: a lexical database for English. Commun. ACM **38**(11), 39–41 (1995)
9. Reimers, N., Gurevych, I.: Sentence-BERT: sentence embeddings using Siamese BERT-networks. In: Proceedings of the 2019 Conference on Empirical Methods in Natural Language Processing. Association for Computational Linguistics (2019)
10. Schneider, P.: App ecosystem out of balance: an empirical analysis of update interdependence between operating system and application software. Frankfurt University Library Johann C. Senckenberg (2020)
11. Schneider, P., Voggenreiter, M., Gulraiz, A., Matthes, F.: Semantic similarity-based clustering of findings from security testing tools. In: Proceedings of the 5th International Conference on Natural Language and Speech Processing (ICNLSP 2022), pp. 134–143. Association for Computational Linguistics, Trento (2022)
12. Tan, X., Peng, X., Pan, S., Zhao, W.: Assessing software quality by program clustering and defect prediction. In: 2011 18th Working Conference on Reverse Engineering, pp. 244–248 (2011)
13. Voggenreiter, M., Schöpp, U.: Using a semantic knowledge base to improve the management of security reports in industrial DevOps projects. In: 2022 IEEE/ACM 44th International Conference on Software Engineering: Software Engineering in Practice (ICSE-SEIP), pp. 309–310 (2022)

Index

M. Abbas (ed.), *Practical Solutions for Diverse Real-World NLP Applications*, Signals and Communication Technology,
https://doi.org/10.1007/978-3-031-44260-5

www.ingramcontent.com/pod-product-compliance
Ingram Content Group UK Ltd.
Pitfield, Milton Keynes, MK11 3LW, UK
UKHW022210051026
15624UKWH00001B/88
9783031442599